A First Course in Finite Element Analysis

A First Course in Finite Element Analysis

Xin-She Yang

University of Cambridge, Cambridge, United Kingdom

Luniver Press

Published in 2007 by Luniver Press
Frome, BA11 6TT, United Kingdom
www.luniver.com

British Library Cataloguing-in-Publication Data
A catalogue record for this book is available from the British Library

ISBN-13: 978-1-905986-08-8
ISBN-10: 1-905986-08-4

Preface

Finite element analysis (FEA) is an essential part of modern scientific computing, and it is a key problem-solving tool for engineers and scientists. Modern engineering design and process modelling require both mathematical analysis and computer simulations. Nowadays, a majority of computer simulations in engineering and computational science are using finite element methods (FEM) and computational fluid dynamcs (CFD) with vast literature.

From my experience in supervising the projects for undergraduates and graduates, I know that many students do not know where to start or how to do finite element analysis and modelling, even though they understand perfectly well the essential concepts of numerical methods. After taking a course of finite element methods for a term or sometimes a year, some students are still confused by the complexity of various formulations and various concepts of finite element methods. This book arises from the teaching and my own learning experience of finite element analysis, and subsequently aims to provide a right level of introduction to the finite element methods for undergraduates and graduates. The topics in finite element analysis are very diverse and the syllabus of numerical methods itself is evolving. Therefore, there is a decision to select the topics and limit the number of chapters so that the book remains concise and yet comprehensive enough to include all the important topics and popular algorithms.

This book endeavors to strike a balance between mathematical and numerical coverage of a wide range of topics in finite element analysis. It strives to provide an introduction, especially for undergraduates and graduates, to finite element analysis and its applications. Topics include advanced calculus, differential equations, vector analysis, calculus of variations, finite difference methods, finite element methods and time-stepping schemes. The book also emphasizes the application of important numerical methods with dozens of worked examples. The applied topics include elasticity, heat transfer, and pattern formation. A few self-explanatory Matlab programs will provide a good start for readers to try some of the methods and/or to apply to their own modelling problems with some modifications. The book can serve as a textbook in finite element analysis, computational mathematics, mathematical modelling, and engineering computations.

Xin-She Yang
Cambridge, 2007

Acknowledgements

First and foremost, I would like to thank my mentors, tutors and colleagues: Profs. A C Fowler and C J Mcdiarmid at Oxford University for introducing me the wonderful world of applied mathematics; Profs. D T Gethin and R W Lewis at Swansea University for introducing me various new methods in the finite element analysis; and Drs J M Lees, C T Morley and G T Parks at Cambridge University for giving me the opportunity to work on the applications of nonlinear finite element analysis.

I thank many of my students at Cambridge University and Swansea University who have directly and/or indirectly tried some parts of this book and gave their valuable suggestions.

I also would like to thank the staff and editors at Luniver Press, especially Prof. Andy Adamatzky and Dr Jon Gelecki, for their kind help and professional editing.

Last but not least, I thank my wife and son for their help and support.

Xin-She Yang

About the Author

Xin-She Yang received his D.Phil in applied mathematics from the University of Oxford. He is currently a research fellow at the University of Cambridge. Dr Yang has published extensively in international journals, book chapters, and conference proceedings. His research interests include asymptotic analysis, bioinspired algorithms, combustion, computational engineering, engineering optimization, finite element analysis, solar eclipses, pattern formation, and scientific computing.

Contents

Chapter 1

Preliminary Mathematics

1.1 Calculus

The preliminary requirements for this book are the pre-calculus foundation mathematics. We assume that the readers are familiar with these preliminaries, therefore, we will only review some of the important concepts of differentiation, integration, Jacobian and multiple integrals.

1.1.1 Differentiations and Integrations

For a known function $y = f(x)$ or a curve as shown in Figure 1.1, the gradient or slope of the curve at any point $P(x, y)$ is defined as

$$\frac{dy}{dx} \equiv \frac{df(x)}{dx} \equiv f'(x) = \lim_{\Delta x \to 0} \frac{f(x + \Delta x) - f(x)}{\Delta x}, \tag{1.1}$$

on the condition that there exists such a limit at P. This gradient or limit is the first derivative of the function $f(x)$ at P. If the limit does not exist at a point P when Δx approaches zero, then we say that the function is non-differentiable at P. By convention, the limit of the infinitesimal change Δx is denoted as the differential dx. Thus, the above definition can also be written as

$$dy = df = \frac{df(x)}{dx}dx = f'(x)dx, \tag{1.2}$$

which can be used to calculate the change in dy caused by the small change of dx. The primed notation $'$ and standard notation $\frac{d}{dx}$ can be used interchangeably, and the choice is purely out of convenience.

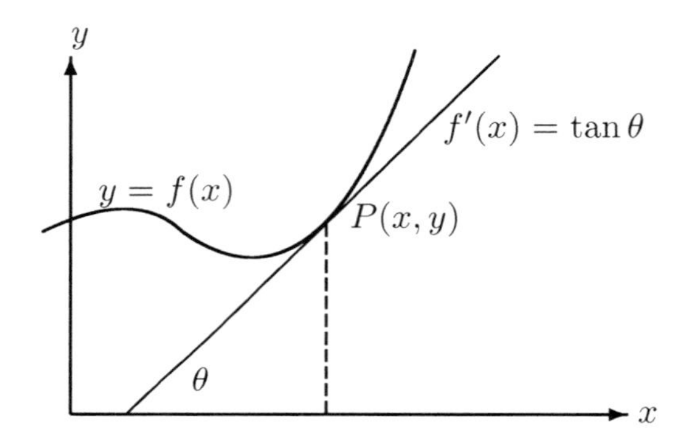

Figure 1.1: Gradient of a function $y = f(x)$

The second derivative of $f(x)$ is defined as the gradient of $f'(x)$, that is to say,

$$\frac{d^2y}{dx^2} \equiv f''(x) = \frac{df(x)}{dx}. \tag{1.3}$$

The higher derivatives can be defined in a similar manner. Thus,

$$\frac{d^3y}{dx^3} \equiv f'''(x) = \frac{df''(x)}{dx}, \quad ..., \quad \frac{d^ny}{dx^n} \equiv f^{(n)} = \frac{df^{(n-1)}}{dx}. \tag{1.4}$$

Differentiation Rules

If a more complicated function $f(x)$ can be written as a product of two simpler functions $u(x)$ and $v(x)$, we can derive a differentiation rule using the definition from the first principles. We have

$$\frac{df}{dx} = \frac{f(x + \Delta x) - f(x)}{\Delta x} = \frac{d[u(x)v(x)]}{dx}$$

$$= \frac{u(x+\Delta x)v(x+\Delta x) - u(x)v(x)}{\Delta x},$$

$$= \lim_{\Delta x \to 0} [u(x+\Delta x)\frac{v(x+\Delta x)-v(x)}{\Delta x} + v(x)\frac{u(x+\Delta x)-u(x)}{\Delta x}]$$

$$= u(x)\frac{dv}{dx} + \frac{du}{dx}v(x), \quad (1.5)$$

which can be written in a compact form using primed notations

$$f'(x) = (uv)' = u'v + uv'. \quad (1.6)$$

If we differentiate this equation again and again, we can get Leibnitz's Theorem for differentiations

$$\frac{d^n(uv)}{dx^n} = u^{(n)}v + nu^{(n-1)}v' + ... + \binom{n}{r} u^{(n-r)}v^{(r)}$$

$$+ ... + uv^{(n)}, \quad (1.7)$$

where the coefficients are the same as the binomial coefficients

$$^nC_r \equiv \binom{n}{r} = \frac{n!}{r!(n-r)!}. \quad (1.8)$$

If a function $f(x)$ [for example, $f(x) = e^{\sin(x)}$] can be written as a function of another function $g(x)$, or $f(x) = f[g(x)]$ [for example, $f(x) = e^{g(x)}$ and $g(x) = \sin(x)$], then we have

$$f'(x) = \lim_{\Delta x \to 0} \frac{\Delta f}{\Delta g}\frac{\Delta g}{\Delta x}, \quad (1.9)$$

which leads to the following chain rule

$$f'(x) = \frac{df}{dg}\frac{dg}{dx}, \quad (1.10)$$

or

$$\{f[g(x)]\}' = f'[g(x)] \cdot g'(x). \quad (1.11)$$

In our example, we have $f'(x) = (e^{\sin(x)})' = e^{\sin(x)} \cos(x)$. If one use $1/v$ instead of v in the equation (1.6) and $(1/v)' = -v'/v^2$, we have the following differentiation rule for quotients:

$$(\frac{u}{v})' = \frac{u'v - uv'}{v^2}. \quad (1.12)$$

The derivatives of various functions are listed in Table 1.1.

Table 1.1: First Derivatives

$f(x)$	$f'(x)$
x^n	nx^{n-1}
e^x	e^x
$a^x (a > 0)$	$a^x \ln a$
$\ln x$	$\frac{1}{x}$
$\log_a x$	$\frac{1}{x \ln a}$
$\sinh x$	$\cosh x$
$\cosh x$	$\sinh x$
$\tan x$	$\sec^2 x$
$\sin^{-1} x$	$\frac{1}{\sqrt{1-x^2}}$
$\cos^{-1} x$	$-\frac{1}{\sqrt{1-x^2}}$
$\tan^{-1} x$	$\frac{1}{1+x^2}$

Implicit Differentiation

The above differentiation rules still apply in the case when there is no simple explicit function form $y = f(x)$ as a function of x only. For example,

$$y^2 - \sin(x)e^y = 0. \tag{1.13}$$

In this case, we can differentiate the equation term by term with respect to x so that we can obtain the derivative dy/dx which is in general a function of both x and y. In our example, we have

$$2y\frac{dy}{dx} - \cos(x)e^y - \sin(x)e^y\frac{dy}{dx} = 0, \tag{1.14}$$

which leads to

$$\frac{dy}{dx} = \frac{\cos(x)e^y}{2y - \sin(x)e^y}. \tag{1.15}$$

Integration

Integration can be viewed as the inverse of differentiation. The integration $F(x)$ of a function $f(x)$ satisfies

$$\frac{dF(x)}{dx} = f(x), \tag{1.16}$$

or

$$F(x) = \int_{x_0}^{x} f(\xi)d\xi, \tag{1.17}$$

where $f(x)$ is called the integrand, and the integration starts from x_0 (arbitrary) to x. In order to avoid any potential confusion, it is conventional to use a dummy variable (say, ξ) in the integrand. As we know, the geometrical meaning of the first derivative is the gradient of the function $f(x)$ at a point P, the geometrical representation of an integral $\int_a^b f(\xi)d\xi$ (with lower integration limit a and upper integration limit b) is the area under the curve $f(x)$ enclosed by x-axis in the region $x \in [a, b]$. In this case, the integral is called a definite integral as the limits are given. For the definite integral, we have

$$\int_a^b f(x)dx = \int_{x_0}^b f(x)dx - \int_{x_0}^a f(x)dx = F(b) - F(a). \tag{1.18}$$

The difference $F(b) - F(a)$ is often written in a compact form $F|_a^b \equiv F(b) - F(a)$. As $F'(x) = f(x)$, we can also write the above equation as

$$\int_a^b f(x)dx = \int_a^b F'(x)dx = F(b) - F(a). \tag{1.19}$$

Since the lower limit x_0 is arbitrary, the change or shift of the lower limit will lead to an arbitrary constant c. When the lower limit is not explicitly given, the integral is called an indefinite integral

$$\int f(x)dx = F(x) + c, \tag{1.20}$$

where c is the constant of integration.

The integrals of some of the common functions are listed in Table 1.2.

Table 1.2: Integrals

$f(x)$	$\int f(x)dx$
$x^n (n \neq -1)$	$\frac{x^{n+1}}{n+1}$
$\frac{1}{x}$	$\ln \lvert x \rvert$
e^x	e^x
$\frac{1}{a^2+x^2}$	$\frac{1}{a} \tan^{-1} \frac{x}{a}$
$\frac{1}{a^2-x^2}$	$\frac{1}{2a} \ln \frac{a+x}{a-x}$
$\frac{1}{x^2-a^2}$	$\frac{1}{2a} \ln \frac{x-a}{x+a}$
$\frac{1}{\sqrt{a^2-x^2}}$	$\sin^{-1} \frac{x}{a}$
$\frac{1}{\sqrt{x^2+a^2}}$	$\ln(x + \sqrt{x^2 + a^2})$ [or $\sinh^{-1} \frac{x}{a}$]
$\frac{1}{\sqrt{x^2-a^2}}$	$\ln(x + \sqrt{x^2 - a^2})$ [or $\cosh^{-1} \frac{x}{a}$]
$\sinh x$	$\cosh x$
$\cosh x$	$\sinh x$
$\tanh x$	$\ln \cosh x$

Integration by Parts

From the differentiation rule $(uv)' = uv' + u'v$, we have

$$uv' = (uv)' - u'v. \tag{1.21}$$

Integrating both sides, we have

$$\int u \frac{dv}{dx} dx = uv - \int \frac{du}{dx} v dx, \tag{1.22}$$

in the indefinite form. It can also be written in the definite form as

$$\int_a^b u \frac{dv}{dx} dx = [uv]\Big|_a^b + \int_a^b v \frac{du}{dx} dx. \tag{1.23}$$

The integration by parts is a very powerful method for evaluating integrals. Many complicated integrands can be rewritten as a product of two simpler functions so that their integrals can easily obtained using integration by parts.

◇ **Example 1.1:** The integral of $I = \int x \ln x \, dx$ can be obtained by setting $v' = x$ and $u = \ln x$. Hence, $v = \frac{x^2}{2}$ and $u' = \frac{1}{x}$. We now have

$$I = \int x \ln x dx = \frac{x^2 \ln x}{2} - \int \frac{x^2}{2} \frac{1}{x} dx$$

$$= \frac{x^2 \ln x}{2} - \frac{x^2}{4}.$$

◇

Other important methods of integration include the substitution and reduction methods. Readers can refer any book that is dedicated to advanced calculus.

Taylor Series and Power Series

From

$$\int_a^b f(x) dx = F(b) - F(a), \tag{1.24}$$

and $\frac{dF}{dx} = F' = f(x)$, we have

$$\int_{x_0}^{x_0+h} f'(x) dx = f(x_0 + h) - f(x_0), \tag{1.25}$$

which means that

$$f(x_0 + h) = f(x_0) + \int_{x_0}^{x_0+h} f'(x) dx. \tag{1.26}$$

If h is not too large or $f'(x)$ does not vary dramatically, we can approximate the integral as

$$\int_{x_0}^{x_0} f'(x) dx \approx f'(x_0) h. \tag{1.27}$$

Thus, we have the first-order approximation to $f(x_0 + h)$

$$f(x_0 + h) \approx f(x_0) + h f'(x_0). \tag{1.28}$$

This is equivalent to say, any change from x_0 to x_0+h is approximated by a linear term $hf'(x_0)$. If we repeat the procedure for $f'(x)$, we have

$$f'(x_0 + h) \approx f'(x_0) + hf''(x_0), \tag{1.29}$$

which is a better approximation than $f'(x_0 + h) \approx f'(x_0)$. Following the same procedure for higher order derivatives, we can reach the n-th order approximation

$$f(x_0 + h) = f(x_0) + hf'(x_0) + \frac{h^2}{2!}f''(x_0) + \frac{h^3}{3!}f'''(x_0)$$

$$+ ... + \frac{h^n}{n!}f^{(n)}(x_0) + R_{n+1}(h), \tag{1.30}$$

where $R_{n+1}(h)$ is the error of this approximation and the notation means that the error is about the same order as $n+1$-th term in the series. This is the well-known Taylor theorem and it has many applications. In deriving this formula, we have implicitly assumed that all the derivatives $f'(x), f''(x), ..., f^{(n)}(x)$ exist. In almost all the applications we meet, this is indeed the case. For example, $\sin(x)$ and e^x, all the orders of the derivatives exist. If we continue the process to infinity, we then reach the infinite power series and the error $\lim_{n\to\infty} R_{n+1} \to 0$ if the series converges. The end results are the Maclaurin series. For example,

$$e^x = 1 + x + \frac{x^2}{2!} + ... + \frac{x^n}{n!} + ..., \quad (x \in \mathcal{R}), \tag{1.31}$$

$$\sin x = x - \frac{x^3}{3!} + \frac{x^5}{5!} - ..., \quad (x \in \mathcal{R}), \tag{1.32}$$

$$\cos x = 1 - \frac{x^2}{2!} + \frac{x^4}{4!} - ..., \quad (x \in \mathcal{R}), \tag{1.33}$$

and

$$\ln(1 + x) = x - \frac{x^2}{2} + \frac{x^3}{3} - \frac{x^4}{4} + \frac{x^5}{5} - ..., \quad x \in (-1, 1]. \tag{1.34}$$

1.1.2 Partial Differentiation

The derivative defined earlier is for function $f(x)$ which has only one independent variable x, and the gradient will generally depend on the location x. For functions $f(x, y)$ of two variables x and y, their gradient will depend on both x and y in general. In addition, the gradient or rate of change will also depend on the direction (along x-axis or y-axis or any other directions). For example, the function $f(x, y) = x(y - 1)$ has different gradients at $(0, 0)$ along x-axis and y-axis. The gradients along the positive x- and y- directions are called the partial derivatives respect to x and y, respectively. They are denoted as $\frac{\partial f}{\partial x}$ and $\frac{\partial f}{\partial y}$, respectively.

The partial derivative of $f(x, y)$ with respect to x can be calculated assuming that y =constant. Thus, we have

$$\frac{\partial f(x,y)}{\partial x} \equiv f_x \equiv \frac{\partial f}{\partial x}|_y$$

$$= \lim_{\Delta x \to 0, y=const} \frac{f(x + \Delta x, y) - f(x, y)}{\Delta x}. \tag{1.35}$$

Similarly, we have

$$\frac{\partial f(x,y)}{\partial y} \equiv f_y \equiv \frac{\partial f}{\partial y}|_x$$

$$= \lim_{\Delta y \to 0, x=const} \frac{f(x, y + \Delta y) - f(x, y)}{\Delta y}. \tag{1.36}$$

The notation $\frac{\partial}{\partial x}|_y$ emphasizes the fact that y is held constant. The subscript notation f_x (or f_y) emphasizes the derivative is carried out with respect to x (or y). Mathematicians like to use the subscript forms as they are simpler notations and can be easily generalized. For example,

$$f_{xx} = \frac{\partial^2 f}{\partial x^2}, \qquad f_{xy} = \frac{\partial^2 f}{\partial x \partial y}. \tag{1.37}$$

Since $\Delta x \Delta y = \Delta y \Delta x$, we have $f_{xy} = f_{yx}$.

For any small change $\Delta f = f(x+\Delta x, y+\Delta y) - f(x, y)$ due to Δx and Δy, the total infinitesimal change df can be written as

$$df = \frac{\partial f}{\partial x} dx + \frac{\partial f}{\partial y} dy. \tag{1.38}$$

If x and y are functions of another independent variable ξ, then the above equation leads to the following chain rule

$$\frac{df}{d\xi} = \frac{\partial f}{\partial x}\frac{dx}{d\xi} + \frac{\partial f}{\partial y}\frac{dy}{d\xi}, \tag{1.39}$$

which is very useful in calculating the derivatives in parametric form or for change of variables. If a complicated function $f(x)$ can be written in terms of simpler functions u and v so that $f(x) = g(x, u, v)$ where $u(x)$ and $v(x)$ are known functions of x, then we have the generalized chain rule

$$\frac{dg}{dx} = \frac{\partial g}{\partial x} + \frac{\partial g}{\partial u}\frac{du}{dx} + \frac{\partial g}{\partial v}\frac{dv}{dx}. \tag{1.40}$$

The extension to functions of more than two variables is straightforward. For a function $p(x, y, z, t)$ such as the pressure in a fluid, we have the total differential as

$$df = \frac{\partial p}{\partial t} dt + \frac{\partial p}{\partial x} dx + \frac{\partial p}{\partial y} dy + \frac{\partial p}{\partial z} dz. \tag{1.41}$$

When differentiating an integral

$$\Phi(x) = \int_a^b \phi(x, y) dy, \tag{1.42}$$

with fixed integration limits a and b, we have

$$\frac{\partial \Phi(x)}{\partial x} = \int_a^b \frac{\partial \phi(x, y)}{\partial x} dy. \tag{1.43}$$

When differentiating the integrals with the limits being functions of x,

$$I(x) = \int_{v(x)}^{u(x)} \psi(x, \tau) d\tau = \Psi[x, u(x)] - \Psi[x, v(x)], \tag{1.44}$$

the following formula is useful:

$$\frac{dI}{dx} = \int_{v(x)}^{u(x)} \frac{\partial \psi}{\partial x} d\tau + [\psi(x, u(x))\frac{du}{dx} - \psi(x, v(x))\frac{dv}{dx}]. \quad (1.45)$$

This formula can be derived using the chain rule

$$\frac{dI}{dx} = \frac{\partial I}{\partial x} + \frac{\partial I}{\partial u}\frac{du}{dx} + \frac{\partial I}{\partial v}\frac{dv}{dx}, \quad (1.46)$$

where $\frac{\partial I}{\partial u} = \psi(x, u(x))$ and $\frac{\partial I}{\partial v} = -\psi(x, v(x))$.

1.1.3 Multiple Integrals

As the integration of a function $f(x)$ corresponds to the area enclosed under the function between integration limits, this can extend to the double integral and multiple integrals. For a function $f(x, y)$, the double integral is defined as

$$F = \int_{\Omega} f(x, y) dA, \quad (1.47)$$

where dA is the infinitesimal element of the area, and Ω is the region for integration. The simplest form of dA is $dA = dxdy$ in Cartesian coordinates. In order to emphasize the double integral in this case, the integral is often written as

$$I = \iint_{\Omega} f(x, y) dxdy. \quad (1.48)$$

◇ **Example 1.2:** The area moment of inertia of a thin rectangular plate, with the width a and the depth b, is defined by

$$I = \iint_{\Omega} y^2 dS = \iint_{\Omega} y^2 dxdy.$$

he plate can be divided into four equal parts, and we have

$$I = 4\int_0^{a/2} [\int_0^{b/2} y^2 dy] dx = 4\int_0^{a/2} \frac{1}{3}(\frac{b}{2})^3 dx$$

$$= \frac{b^3}{6}\int_0^{a/2} dx = \frac{ab^3}{12}.$$

◇

1.1.4 Jacobian

Sometimes it is necessary to change variables when evaluating an integral. For a simple one-dimensional integral, the change of variables from x to a new variable v (say) leads to $x = x(v)$. This is relatively simple as $dx = \frac{dx}{dv}dv$, and we have

$$\int_{x_a}^{x_b} f(x)dx = \int_a^b f(x(v))\frac{dx}{dv}dv, \qquad (1.49)$$

where the integration limits change so that $x(a) = x_a$ and $x(b) = x_b$. Here the extra factor dx/dv in the integrand is referred to as the Jacobian.

For a double integral, it is more complicated. Assuming $x = x(\xi, \eta), y = y(\xi, \eta)$, we have

$$\iint f(x,y)dxdy = \iint f(\xi\eta)|J|d\xi d\eta, \qquad (1.50)$$

where J is the Jacobian. That is

$$J \equiv \frac{\partial(x,y)}{\partial(\xi,\eta)}$$

$$= \begin{vmatrix} \frac{\partial x}{\partial \xi} & \frac{\partial x}{\partial \eta} \\ \frac{\partial y}{\partial \xi} & \frac{\partial y}{\partial \eta} \end{vmatrix} = \begin{vmatrix} \frac{\partial x}{\partial \xi} & \frac{\partial y}{\partial \xi} \\ \frac{\partial x}{\partial \eta} & \frac{\partial y}{\partial \eta} \end{vmatrix}. \qquad (1.51)$$

The notation $\partial(x,y)/\partial(\xi,\eta)$ is just a useful shorthand. This is equivalent to say that the change of the infinitesimal area $dA = dxdy$ becomes

$$dxdy = |\frac{\partial(x,y)}{\partial(\xi,\eta)}|d\xi d\eta = |\frac{\partial x}{\partial \xi}\frac{\partial y}{\partial \eta} - \frac{\partial x}{\partial \eta}\frac{\partial y}{\partial \xi}|d\xi d\eta. \qquad (1.52)$$

◇ **Example 1.3:** When transforming from (x,y) to polar coordinates (r,θ), we have the following relationships

$$x = r\cos\theta, \qquad y = r\sin\theta.$$

Thus, the Jacobian is

$$J = \frac{\partial(x,y)}{\partial(r,\theta)} = \frac{\partial x}{\partial r}\frac{\partial y}{\partial \theta} - \frac{\partial x}{\partial \theta}\frac{\partial y}{\partial r}$$

$$= \cos\theta \times r\cos\theta - (-r\sin\theta)\times\sin\theta = r[\cos^2\theta + \sin^2\theta] = r.$$

Thus, an integral in (x, y) will be transformed into

$$\iint \phi(x,y)dxdy = \iint \phi(r\cos\theta, r\sin\theta)rdrd\theta.$$

◇

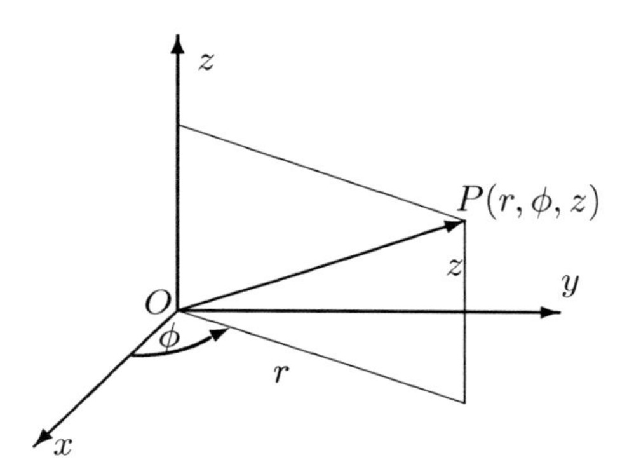

Figure 1.2: Cylindrical polar coordinates.

In a similar fashion, the change of variables in triple integrals gives

$$V = \iiint_\Omega \phi(x,y,z)dxdydz = \iint_\omega \psi(\xi,\eta,\zeta)|J|d\xi d\eta d\zeta, \quad (1.53)$$

and

$$J \equiv \frac{\partial(x,y,z)}{\partial(\xi,\eta,\zeta)} = \begin{vmatrix} \frac{\partial x}{\partial \xi} & \frac{\partial y}{\partial \xi} & \frac{\partial z}{\partial \xi} \\ \frac{\partial x}{\partial \eta} & \frac{\partial y}{\partial \eta} & \frac{\partial z}{\partial \eta} \\ \frac{\partial x}{\partial \zeta} & \frac{\partial y}{\partial \zeta} & \frac{\partial z}{\partial \zeta} \end{vmatrix}. \quad (1.54)$$

For cylindrical polar coordinates (r, ϕ, z) as shown in Figure 1.2, we have

$$x = r\cos\phi, \qquad y = r\sin\phi, \qquad z = z. \quad (1.55)$$

The Jacobian is therefore

$$J = \frac{\partial(x,y,z)}{\partial(r,\phi,z)} = \begin{vmatrix} \cos\phi & \sin\phi & 0 \\ -r\sin\phi & r\cos\phi & 0 \\ 0 & 0 & 1 \end{vmatrix} = r. \quad (1.56)$$

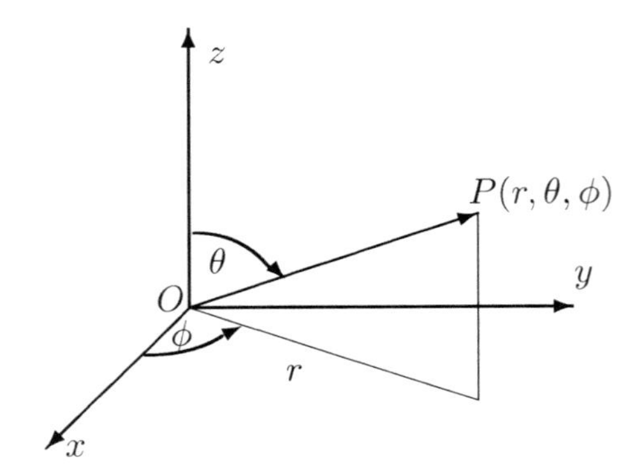

Figure 1.3: Spherical polar coordinates.

For spherical polar coordinates (r, θ, ϕ) as shown in Figure 1.3, where θ is the zenithal angle between the z-axis and the position vector $\mathbf{r}$, and ϕ is the azimuthal angle, we have

$$x = r \sin\theta \cos\phi, \qquad y = r \sin\theta \sin\phi, \qquad z = r \cos\theta. \quad (1.57)$$

Therefore, the Jacobian is

$$J = \begin{vmatrix} \sin\theta\cos\phi & \sin\theta\sin\phi & \cos\theta \\ r\cos\theta\cos\phi & r\cos\theta\sin\phi & -r\sin\theta \\ -r\sin\theta\sin\phi & r\sin\theta\cos\phi & 0 \end{vmatrix} = r^2 \sin\theta. \quad (1.58)$$

Thus, the volume element change in the spherical system is

$$dxdydz = r^2 \sin\theta dr d\theta d\phi. \quad (1.59)$$

1.1.5 Numerical Integration

An interesting feature about the differentiations and integrations is that you can get the explicit expressions of derivatives of most functions and complicated expressions if they exist, while it is very difficult and sometimes impossible to express an integral in an explicit form, even for seemingly simple integrands. For example, the error function, widely used in engineering and sciences, is defined as

$$\text{erf}(x) = \frac{2}{\sqrt{\pi}} \int_0^x e^{-t^2} dt. \tag{1.60}$$

The integration of this simple integrand $\exp(-t^2)$ does not leads to any simple explicit expression, which is why it is often written as erf(), referred to as the error function. If we pick up a mathematical handbook, we know that

$$\text{erf}(0) = 0, \qquad \text{erf}(\infty) = 1, \tag{1.61}$$

while

$$\text{erf}(0.5) \approx 0.52049, \qquad \text{erf}(1) \approx 0.84270. \tag{1.62}$$

If we want to calculate such integrals, numerical integration is the best alternative.

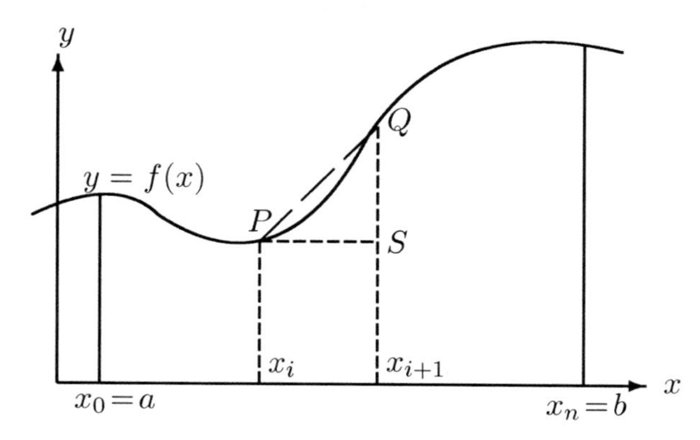

Figure 1.4: Integration of a function $y = f(x)$

Trapezium Rule

Now if we want to numerically evaluate the following integral

$$\mathcal{I} = \int_a^b f(x) dx, \tag{1.63}$$

where a and b are fixed and finite. We know that the value of the integral is exactly the total area under the curve $y = f(x)$ between a and b. As both the integral and the area can be considered as the sum of the values over many small intervals, the simplest way for evaluating such numerical integration is to divide up the integral interval into n equal small sections and split the area into n thin strips so that $h \equiv \Delta x = (b-a)/n$, $x_0 = a$ and $x_i = ih + a (i = 1, 2, ..., n)$. The values of the functions at the dividing points x_i are denoted as $y_i = f(x_i)$, and the value at the midpoint between x_i and x_{i+1} is labelled as $y_{i+1/2} = f_{i+1/2}$

$$y_{i+1/2} = f(x_{i+1/2}) = f_{i+1/2}, \qquad x_{i+1/2} = \frac{x_i + x_{i+1}}{2}. \tag{1.64}$$

The accuracy of such approximations depends on the number n and the way to approximate the curve in each interval. Figure 1.4 shows such an interval $[x_i, x_{i+1}]$ which is exaggerated in the figure for clarity. The curve segment between P and Q is approximated by a straight line with a slope

$$\frac{\Delta y}{\Delta x} = \frac{f(x_{i+1}) - f(x_i)}{h}, \tag{1.65}$$

which approaches $f'(x_{i+1/2})$ at the midpoint point when $h \to 0$.

The trapezium (formed by P, Q, x_{i+1}, and x_i) is a better approximation than the rectangle (P, S, x_{i+1} and x_i) because the former has an area

$$A_i = \frac{f(x_i) + f(x_{i+1})}{2} h, \tag{1.66}$$

which is close to the area

$$\mathcal{I}_i = \int_{x_i}^{x_{i+1}} f(x) dx, \tag{1.67}$$

under the curve in the small interval x_i and x_{i+1}. If we use the area A_i to approximate $\mathcal{I}_i$, we have the trapezium rule of numerical integration. Thus, the integral is simply the sum of all these small trapeziums, and we have

$$\mathcal{I} \approx \frac{h}{2}[f_0 + 2(f_1 + f_2 + ... + f_{n-1}) + f_n]$$

$$= h[f_1 + f_2 + ... + f_{n-1} + \frac{(f_0 + f_n)}{2}]. \tag{1.68}$$

From the Taylor series (1.30), we know that

$$\frac{f(x_i) + f(x_{i+1})}{2}$$

$$\approx \frac{1}{2}\Big\{[f(x_{i+1/2}) - \frac{h}{2}f'(x_{i+1/2}) + \frac{1}{2!}(\frac{h}{2})^2 f''(x_{i+1/2})]$$

$$+[f(x_{i+1/2}) + \frac{h}{2}f'(x_{i+1/2}) + \frac{1}{2!}(\frac{h}{2})^2]f''(x_{i+1/2})\Big\}$$

$$= f(x_{i+1/2}) + \frac{h^2}{8}f''(x_{i+1/2}). \tag{1.69}$$

where $O(h^2 f'')$ means that the value is the order of $h^2 f''$, or $O(h^2) = Kh^2 f''$ where K is a constant. Therefore, the error of the estimate of $\mathcal{I}$ is $h \times O(h^2 f'') = O(h^3 f'')$.

Now let us briefly introduce the order notations. Loosely speaking, for two functions $f(x)$ and $g(x)$, if

$$\lim_{x \to x_0} \frac{f(x)}{g(x)} \to K, \qquad x \to x_0, \tag{1.70}$$

where K is a finite, non-zero limit, we write

$$f = O(g). \tag{1.71}$$

The big O notation means that f is asymptotically equivalent to the order of $g(x)$. If the limit is unity or $K = 1$, we say $f(x)$ is order of $g(x)$. In this special case, we write

$$f \sim g, \tag{1.72}$$

which is equivalent to $f/g \to 1$ and $g/f \to 1$ as $x \to x_0$. Obviously, x_0 can be any value, including 0 and ∞. The notation $\sim$ does not necessarily mean $\approx$ in general, though they might give the same results, especially in the case when $x \to 0$ [for example, $\sin x \sim x$ and $\sin x \approx x$ if $x \to 0$].

When we say f is order of 100 (or $f \sim 100$), this does not mean $f \approx 100$, but it can mean that f is between about 50 to

150. The small o notation is used if the limit tends to 0. That is

$$\frac{f}{g} \to 0, \qquad x \to x_0, \tag{1.73}$$

or

$$f = o(g). \tag{1.74}$$

If $g > 0$, $f = o(g)$ is equivalent to $f \ll g$. For example, for $\forall x \in \mathcal{R}$, we have $e^x \approx 1 + x + O(x^2) \approx 1 + x + \frac{x^2}{2} + o(x)$.

Another classical example is Stirling's asymptotic series for factorials

$$n! \sim \sqrt{2\pi n}(\frac{n}{e})^n(1 + \frac{1}{12n} + \frac{1}{288n^2} - \frac{139}{51480n^3} - ...). \tag{1.75}$$

This is a good example of asymptotic series. For standard power expansions, the error $R_k(h^k) \to 0$, but for an asymptotic series, the error of the truncated series R_k decreases and gets smaller compared with the leading term [here $\sqrt{2\pi n}(n/e)^n$]. However, R_n does not necessarily tend to zero. In fact, $R_2 = \frac{1}{12n} \cdot \sqrt{2\pi n}(n/e)^n$ is still very large as $R_2 \to \infty$ if $n \gg 1$. For example, for $n = 100$, we have $n! = 9.3326 \times 10^{157}$, while the leading approximation is $\sqrt{2\pi n}(n/e)^n = 9.3248 \times 10^{157}$. The difference between these two values is 7.7740×10^{154}, which is still very large, though three orders smaller than the leading approximation.

Simpson's Rule

The trapezium rule introduced earlier is just one of the simple and popular schemes for numerical integration with the error of $O(h^3 f'')$. If we want higher accuracy, we can either reduce h or use a better approximation for $f(x)$). A small h means a large n, which implies that we have to do the sum of many small sections, and it may increase the computational time. On the other hand, we can use higher order approximations for the curve. Instead of using straight lines or linear approximations for curve segments, we can use parabolas or quadratic approximations. For any consecutive three points x_{i-1}, x_i and x_{i+1}, we can construct a parabola in the form

$$f(x_i+t)=f_i+\alpha t+\beta t^2, \qquad t\in[-h,h]. \tag{1.76}$$

As this parabola must go through the three known points (x_{i-1}, f_{i-1}) at $t=-h$, (x_i,f_i) at $t=0$ and x_{i+1}, f_{i+1} at $t=h$, we have the following equations for α and β

$$f_{i-1}=f_i-\alpha h+\beta h^2, \tag{1.77}$$

and

$$f_{i+1}=f_i+\alpha h+\beta h^2, \tag{1.78}$$

which lead to

$$\alpha=\frac{f_{i+1}-f_{i-1}}{2h}, \qquad \beta=\frac{f_{i-1}-2f_i+f_{i+1}}{h^2}. \tag{1.79}$$

We will see in later chapters that α is the centred approximation for the first derivative f_i' and β is the central difference scheme for the second derivative f_i''. Therefore, the integral from x_{i-1} to x_{i+1} can be approximated by

$$\mathcal{I}_i=\int_{x_{i-1}}^{x_{i+1}} f(x)dx \approx \int_{-h}^{h}[f_i+\alpha t+\beta t^2]dt$$

$$=\frac{h}{3}[f_{i-1}+4f_i+f_{i+1}], \tag{1.80}$$

where we have substituted the expressions for α and β. To ensure the whole interval $[a,b]$ can be divided up to form three-point approximations without any point left out, n must be even. Therefore, the estimate of the integral becomes

$$\mathcal{I}\approx\frac{h}{3}[f_0+4(f_1+f_3+...+f_{n-1})$$

$$+2(f_2+f_4+...+f_{n-2})+f_n], \tag{1.81}$$

which is the standard Simpson's rule.

As the approximation for the function $f(x)$ is quadratic, an order higher than the linear form, the error estimate of Simpson's rule is thus $O(h^4)$ or $O(h^4 f'''')$ to be more specific. There are many variations of Simpson's rule with higher order accuracies such as $O(h^5 f^{(4)})$ and $O(h^7 f^{(6)})$.

1.1.6 Gaussian Integration

To get higher-order accuracy, we can use polynomials to construct various integration schemes. However, there is an easier way to do this. That is to use the Gauss-Legendre intergration or simply Gaussian integration. Since any integral $\mathcal{I}$ with integration limits a and b can be transformed to an integral with limits -1 and $+1$ by using

$$\zeta = \frac{2x}{(b-a)} - 1, \tag{1.82}$$

so that

$$\mathcal{I} = \int_a^b g(x)dx = \frac{(b-a)}{2} \int_{-1}^1 f(\zeta)d\zeta, \tag{1.83}$$

where we have used $dx = (b-a)d\zeta/2$. Therefore, we only have to study the integral

$$J = \int_{-1}^1 f(\zeta)d\zeta. \tag{1.84}$$

The n values of the function or n integration points are given by a polynomial of $n-1$ degree. For equal spacing h, this numerical integration technique is often referred to as the Newton-Cotes quadrature

$$J = \int_{-1}^1 f(d\zeta)d\zeta = \sum_{i=1}^n w_i f(\zeta_i), \tag{1.85}$$

where w_i is the weighting coefficient attached to $f(\zeta_i)$. Such integral will have an error of $O(h^n)$. For example, $n=2$ with equal weighting corresponds to the trapezium rule because

$$J = f_{-1} + f_1. \tag{1.86}$$

For the case of $n=3$, we have

$$J = \frac{1}{3}[f_{-1} + 4f_0 + f_1], \tag{1.87}$$

which corresponds to Simpson's rule.

The numerical integration we use so far are carried out at equally-spaced points $x_0, x_1, ..., x_i, ..., x_n$, and these points are fixed *a priori*. There is no particular reason why we should use the equally-spaced points apart from the fact that it is easy and simple. In fact, we can use any sampling points or integration points as we wish to improve the accuracy of the estimate to the integral. If we use n integration points $(\zeta_i, i = 1, 2, ..., n)$ with a polynomial of $2n - 1$ degrees or Legendre polynomial $P_n(x)$, we now have $2n$ unknowns f_i and ζ_i. This means that we can easily construct quadrature formula, often called, Gauss quadrature or Gaussian integration.

Mathematically, we have the Gauss quadrature

$$J = \int_{-1}^{1} f(\zeta) d\zeta = \sum_{i=1}^{n} w_i f(\zeta_i), \tag{1.88}$$

where ζ_i is determined by the zeros of the Legendre polynomial $P_n(\zeta_i) = 0$ and the weighting coefficient is given by

$$w_i = \frac{2}{(1 - \zeta_i^2)[P_n'(\zeta_i)]^2}. \tag{1.89}$$

The error of this quadrature is of order $O(h^{2n})$. The proof of this formulation is out of the scope of this book. Readers can find the proof in more advanced mathematical books.

Briefly speaking, Legendre polynomials are obtained by the following generating function or Rodrigue's formula

$$P_n(x) = \frac{1}{2^n n!} \frac{d^n (n^2 - 1)^n}{dx^n}. \tag{1.90}$$

For example,

$$P_0(x) = 0, \quad P_1(x) = x, \quad P_2(x) = \frac{1}{2}(3x^2 - 1), \tag{1.91}$$

and

$$P_3(x) = \frac{1}{2}(5x^3 - 3x), \quad P_4 = \frac{1}{8}(3 - 30x^2 + 35x^4). \tag{1.92}$$

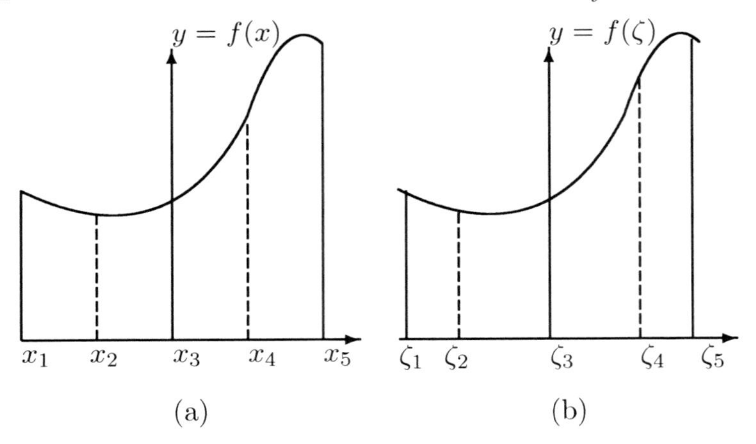

Figure 1.5: Five integration points: a) Equally spaced with $h = 1/4$; and b) Gauss points with $|\zeta_1 - \zeta_3| = |\zeta_5 - \zeta_3| \approx 0.90617$ and $|\zeta_2 - \zeta_3| = |\zeta_4 - \zeta_3| \approx 0.53847$.

For both Newton-Cotes quadrature and Gauss quadrature, Figure 1.5 shows the their difference and similarity.

The computation of locations ζ_i of the Gaussian integration points and weighting coefficients w_i is complicated though straightforward once we know the Legendre polynomials. For example, $n = 2$, we have

$$P_2(\zeta) = \frac{1}{2}(3\zeta^2 - 1) = 0, \tag{1.93}$$

which has two solutions

$$\zeta_{\pm 1} = \pm\sqrt{\frac{1}{3}} \approx \pm 0.5774. \tag{1.94}$$

Since $P_2'(\zeta) = 3\zeta$, we have

$$w_1 = \frac{2}{(1 - \zeta_1^2)(3\zeta_1)^2}$$

$$= \frac{2}{(1 - (\sqrt{1/3})^2) * (3\sqrt{1/3})^2} = 1 = w_{-1}. \tag{1.95}$$

The coefficients w_i and the integration points are usually listed tables for various values of n.

For multiple integrals

$$J = \int_{-1}^{1} \int_{-1}^{1} f(\zeta, \eta) d\zeta d\eta, \tag{1.96}$$

these Gaussian quadrature can easily be extended by evaluating the integral with η being kept constant first, then evaluating the outer integral. We have

$$J = \int_{-1}^{1} [\int_{-1}^{n} f(\zeta, \eta) d\zeta] d\eta$$

$$= \int_{-1}^{1} \sum_{i=1}^{n} w_i f(\zeta_i, \eta) d\eta = \sum_{i=1}^{n} w_i \int_{-1}^{1} f(\zeta_i, \eta) d\eta$$

$$= \sum_{i=1}^{n} \sum_{j=1}^{n} w_i w_j f(\zeta_i, \eta_j), \tag{1.97}$$

where we have used

$$\int_{-1}^{1} f(\zeta_i, \eta) d\eta = \sum_{j=1}^{n} w_j f(\zeta_i, \eta_j). \tag{1.98}$$

◇ **Example 1.4:** To evaluate the integral

$$I = \frac{2}{\sqrt{\pi}} \int_{-1}^{1} e^{-x^2} dx = \int_{-1}^{1} f(x) dx, \quad f(x) = \frac{2}{\sqrt{\pi}} e^{-x^2}.$$

We know that its exact value from (1.60) is

$$I = 2 \operatorname{erf}(1) = 1.685401585899...$$

Let us now estimate it using Simpson's rule for three point integration at $x_{-1} = -1$, $x_0 = 0$ and $x_1 = 1$, and we have

$$I \approx \frac{1}{3}(f_{-1} + 4f_0 + f_1)$$

$$\approx \frac{2}{3\sqrt{\pi}}[e^{-(-1)^2} + 4 \times 1 + e^{-(1)^2}] \approx 1.7812,$$

which differs from its exact value by about 5.6%.

If we use the 3-point Gauss quadrature at $x_{\pm 1} = \pm\sqrt{\frac{3}{5}}$ and $x_0 = 0$ with weighting coefficients $w_{\pm 1} = \frac{5}{9}$ and $w_0 = \frac{8}{9}$, we have

$$I \approx \sum_{i=-1}^{1} w_i f(x_i)$$

$$\approx \frac{2}{\sqrt{\pi}}[\frac{5}{9}e^{-(-\sqrt{3/5})^2} + \frac{8}{9} \times 1 + \frac{5}{9}e^{-(\sqrt{3/5})^2}] \approx 1.6911,$$

which is a better approximation than 1.7812. In fact, the error of Gauss quadrature is just $(1.6911 - 1.6854)/1.6911 \approx 0.3\%$. This higher accuracy is why the Gaussian quadrature rule is so widely used. ◇

For triple integrals and other integrals, the Gauss quadrature can be constructed in a similar way. We will see more numerical techniques in the rest of the book.

1.2 Complex Variables

Although all the quantities are real variables in the physical world, however, it is sometimes easy or even necessary to use complex variables in mathematics and engineering. In fact, the techniques based on complex variables are among the most powerful methods for mathematical analysis and solutions of mathematical models.

1.2.1 Complex Numbers

Mathematically speaking, a complex number z is a generalized set or the order pair of two real numbers (a, b), written in the form of

$$z = a + ib, \qquad i^2 = -1, \qquad a, b \in \mathcal{R}, \tag{1.99}$$

which consists of the real part $\Re(z) = a$ and the imagery part $\Im(z) = b$. It can also be written as the order pair of real numbers using the notation (a, b). The addition and substraction of two complex numbers are defined as

$$(a + ib) \pm (c + id) = (a \pm c) + i(b \pm d). \tag{1.100}$$

The multiplication and division of two complex numbers are in the similar way as expanding polynomials

$$(a + ib) \cdot (c + id) = (ac - bd) + i(ad + bc), \tag{1.101}$$

and

$$\frac{a + ib}{c + id} = \frac{ac + bd}{c^2 + d^2} + i\frac{bc - ad}{c^2 + d^2}. \tag{1.102}$$

Two complex numbers are equal $a + ib = c + id$ if and only if $a = c$ and $b = d$. The complex conjugate or simply conjugate $\bar{z}$ (also z^*) of $z = a + ib$ is defined as

$$\bar{z} = a - ib. \tag{1.103}$$

The order pair (a, b), similar to a vector, implies that a geometrical representation of a complex number $a + ib$ by the point in an ordinary Euclidean plane with x-axis being the real axis and y-axis being the imaginary axis (iy). This plane is called the complex plane. This representation is often called the Argand diagram (see Figure 1.6). The vector representation starts from $(0, 0)$ to the point (a, b). The length of the vector is called the magnitude or modulus or the absolute value of the complex number

$$r = |z| = \sqrt{a^2 + b^2}. \tag{1.104}$$

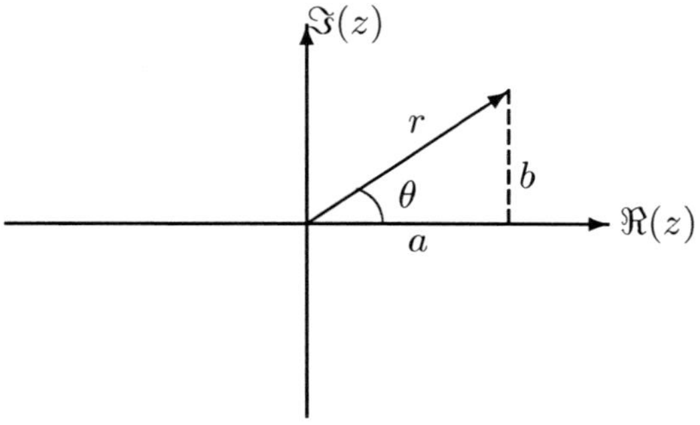

Figure 1.6: Polar representation of a complex number.

The angle θ that the vector makes with the positive real axis is called the argument (see Fig 1.6),

$$\theta = \arg z. \tag{1.105}$$

In fact, we may replace θ by $\theta + 2n\pi$ ($n \in \mathcal{N}$). The value range $-\pi < \theta \leq \pi$ is called the principal argument of z, and it is usually denoted as Arg$z$. In the complex plane, the complex number can be written as

$$z = re^{i\theta} = r\cos(\theta) + ir\sin(\theta). \quad (1.106)$$

This polar form of z and its geometrical representation can result in the Euler's formula which is very useful in the complex analysis

$$e^{i\theta} = \cos(\theta) + i\sin(\theta). \quad (1.107)$$

The Euler formula can be proved using the power series. For any $z \in \mathcal{C}$, we have the power series

$$e^z = 1 + z + \frac{z^2}{2!} + ... + \frac{z^n}{n!} + ..., \quad (1.108)$$

and for a special case $z = i\theta$, we have

$$e^{i\theta} = 1 + i\theta - \frac{\theta^2}{2!} + \frac{i\theta^3}{3!} - ...,$$

$$= (1 - \frac{\theta^2}{2!} + ...) + i(\theta - \frac{\theta^3}{3!} + ...). \quad (1.109)$$

Using the power series

$$\sin\theta = \theta - \frac{\theta^3}{3!} + \frac{\theta^5}{5!} - ..., \quad (1.110)$$

and

$$\cos\theta = 1 - \frac{\theta^2}{2!} + \frac{\theta^4}{4!} - ..., \quad (1.111)$$

we get the well-know Euler's formula or Euler's equation

$$e^{i\theta} = \cos\theta + i\sin\theta. \quad (1.112)$$

For $\theta = \pi$, this leads to a very interesting formula

$$e^{i\pi} + 1 = 0. \quad (1.113)$$

If we replace θ by $-\theta$, the Euler's formula becomes

$$e^{-i\theta} = \cos(-\theta) + i\sin(-\theta) = \cos\theta - i\sin\theta. \tag{1.114}$$

Adding this equation to (1.112), we have

$$e^{i\theta} + e^{-i\theta} = 2\cos\theta, \tag{1.115}$$

or

$$\cos\theta = \frac{e^{i\theta} + e^{-i\theta}}{2}. \tag{1.116}$$

Similarly, by deducting (1.114) from (1.112), we get

$$\sin\theta = \frac{e^{i\theta} - e^{-i\theta}}{2i}. \tag{1.117}$$

For two complex numbers $z_1 = r_1 e^{i\alpha_1}$ and $z_2 = r_2 e^{i\alpha_2}$, it is straightforward to show that

$$z_1 z_2 = r_1 r_2 e^{i(\alpha_1+\alpha_2)}$$

$$= r_1 r_2[\cos(\alpha_1 + \alpha_2) + i\sin(\alpha_1 + \alpha_2)], \tag{1.118}$$

which can easily be extended using $\alpha_1 = \alpha_2 = \theta$ and applying n times to get the well-known de Moivre's formula

$$[\cos(\theta) + i\sin(\theta)]^n = \cos(n\theta) + i\sin(n\theta). \tag{1.119}$$

1.2.2 Analytic Functions

Analytic Functions

Any function of real variables can be extended to the function of complex variables in the same form while treating the real numbers x as $x + i0$. For example, $f(x) = x^2, x \in \mathcal{R}$ becomes $f(z) = z^2, z \in \mathcal{C}$. Any complex function $f(z)$ can be written as

$$f(z) = f(x + iy) = \Re(f(z)) + i\Im(f(z))$$

$$= u(x, y) + iv(x, t), \tag{1.120}$$

where $u(x, y)$ and $v(x, y)$ are real-valued functions of two real variables.

A function $f(z)$ is called analytic at z_0 if $f'(z)$ exists for all z in some $\epsilon-$neighborhood of z_0, that is to say, it is differentiable in some open disk $|z - z_0| < \epsilon$. If $f(z) = u + iv$ is analytic at every point in a domain Ω, then $u(x, y)$ and $v(x, y)$ satisfying the Cauchy-Riemann equations

$$\frac{\partial u}{\partial x} = \frac{\partial v}{\partial y}, \qquad \frac{\partial u}{\partial y} = -\frac{\partial v}{\partial x}. \tag{1.121}$$

Conversely, if u and v of $f(z) = u + iv$ satisfy the Cauchy-Riemann equation at all points in a domain, then the complex function $f(z)$ is analytic in the same domain. For example, the elementary power function $w = z^n, (n > 1)$ is analytic on the whole plane, $w = \rho e^{i\phi}, \qquad z = re^{i\theta}$, then

$$\rho = r^n, \phi = n\theta. \tag{1.122}$$

The logarithm is also an elementary function $w = \ln z$

$$\ln z = \ln |z| + i \arg(z) = \ln r + i(\theta + w\pi k), \tag{1.123}$$

which has infinitely many values, due to the multiple values of θ, with the difference of $2\pi ik(k = 0, \pm1, \pm2, ...)$. If we use the principal argument Arg$z$, then we have the principal logarithm function

$$\text{Ln}(z) = \ln |z| + \text{Arg}z. \tag{1.124}$$

If we differentiate the Cauchy-Riemann equations, we have $\partial^2 u/\partial x\partial y = \partial^2 u/\partial y\partial x$. After some calculations, we can reach the following theorem. For given analytic function $f(z) = u + iv$, then both u and v satisfy the Laplace equations

$$\frac{\partial^2 u}{\partial x^2} + \frac{\partial^2 v}{\partial y^2} = 0, \qquad \frac{\partial^2 v}{\partial x^2} + \frac{\partial^2 v}{\partial y^2} = 0. \tag{1.125}$$

This is to say, both real and imaginary parts of an analytic function are harmonic.

Any analytic function $f(z)$ can be expanded in terms of the Taylor series

$$f(z) = \sum_{k=0}^{\infty} \frac{f^{(k)}}{k!}(z - z_0)^k = \sum_{k=0}^{\infty} \alpha_k (z - z_0)^k. \tag{1.126}$$

This expansion is valid inside the analytic region. However, if the function $f(z)$ has a pole or singularity of order n at $z = z_0$ and it is analytic everywhere except at the pole, we can then expand the function $p(z) = (z - z_0)^n f(z)$ in the standard Taylor expansion. This means that original function $f(z)$ can be written as a power series

$$f(z) = \frac{\alpha_{-n}}{(z - z_0)^n} + ... + \frac{\alpha_{-1}}{(z - z_0)}$$

$$+\alpha_0(z - z_0) + ... + \alpha_k(z - z_0)^k + ..., \tag{1.127}$$

which is called a Laurent series, and it is an extension of the Taylor series. In this series, it is often assumed that $\alpha_{-n} \neq 0$. The terms with the inverse powers $a_{-n}/(z-z_0)^n + ... + a_{-1}/(z-z_0)$ are called the principal part of the series, while the usual terms $a_0(z - z_0) + ... + \alpha_k(z - z_0)^k + ...$ are called the analytic part.

Furthermore, the most important coefficient is probably α_{-1} which is called the residue of $f(z)$ at the pole $z = z_0$. In general, the Laurent series can be written as

$$f(z) = \sum_{k=-n}^{\infty} \alpha_k(z - z_0)^k, \tag{1.128}$$

where n may be extended to include an infinite number of terms $n \to -\infty$.

1.2.3 Complex Integrals

Given a function $f(z)$ that is continuous on a piecewise smooth curve Γ, then the integral over Γ, $\int_\Gamma f(z)dz$, is called a contour or line integral of $f(z)$. This integral has similar properties as the real integral

$$\int_\Gamma [\alpha f(z) + \beta g(z)] dz = \alpha \int_\Gamma f(z) dz + \beta \int_\Gamma g(z) dz. \quad (1.129)$$

If $F(z)$ is analytic and $F'(z) = f(z)$ is continuous along a curve Γ, then

$$\int_a^b f(z) dz = F[z(b)] - F[z(a)]. \quad (1.130)$$

Cauchy's Integral Theorem

We say a path is simply closed if its end points and initial points coincide and the curve does not cross itself. For an analytic function $f(z) = u(x, y) + iv(x, y)$, the integral on a simply closed path

$$I = \int_\Gamma f(z) dz = \int_\Gamma (u + iv)(dx + idy)]$$

$$= \int_\Gamma (udx - vdy) + i \int_\Gamma (vdx + udy). \quad (1.131)$$

By using the Green theorem (see Chapter 2), we have

$$I = \int_\Omega (-\frac{\partial u}{\partial y} - \frac{\partial v}{\partial x}) dxdy + i \int_\Omega (\frac{\partial u}{\partial x} - \frac{\partial v}{\partial y}) dxdy. \quad (1.132)$$

From the Cauchy-Riemann equations, we know that both integrals are zero. Thus, we have Cauchy's Integral Theorem, which states that the integral of any analytic function $f(z)$ on a simply closed path Γ in a simply connected domain Ω is zero. That is

$$\int_\Gamma f(z) dz = 0.$$

This theorem is very important as it has interesting consequences. If the close path is decomposed into two paths with reverse directions Γ_1 and Γ_2, then Γ_1 and $-\Gamma_2$ form a close path, which leads to

$$\int_{\Gamma_1} f(z) dz = \int_{\Gamma_2} f(z) dz. \quad (1.133)$$

That is to say that the integrals over any curve between two points are independent of path. This property becomes very useful for evaluation of integrals. For an analytic function with a pole, we can make the contour Γ sufficiently small to enclose just around the pole, and this makes the calculation of the integral much easier in some cases.

For the integral of $p(z) = f(z)/(z - z_0)$ over any simply closed path Γ enclosing a point z_0 in the domain Ω,

$$I = \int_\Gamma p(z)dz, \tag{1.134}$$

we can use the Laurent series for $p(z)$

$$p(z) = \frac{\alpha_{-1}}{(z - z_0)} + \alpha_0(z - z_0) + ... + \alpha_k(z - z_0)^k + ..., \tag{1.135}$$

so that the expansion can be integrated term by term around a path. The only non-zero contribution over a small circular contour is the residue α_{-1}. We have

$$I = \int_\Gamma p(z)dz = 2\pi i\alpha_{-1} = 2\pi i \,\mathrm{Res}[p(z)]\Big|_{z_0}, \tag{1.136}$$

which can be written in terms of $f(z)$ as

$$\frac{1}{2\pi i}\oint_\Gamma \frac{f(z)}{z - z_0}dz = f(z_0). \tag{1.137}$$

Similarly, this can be extended for higher derivatives, and we have

$$\oint_\Gamma \frac{f(z)}{(z - z_0)^{n+1}}dz = \frac{2\pi i f^{(n)}(z_0)}{n!}.$$

Residue Theorem

For any analytic $f(z)$ function in a domain Ω except isolated singularities at finite points $z_1, z_2, ..., z_N$, the residue theorem states

$$\oint_\Gamma f(z)dz = 2\pi i \sum_{k=1}^{N} \mathrm{Res} f(z)|_{z_k},$$

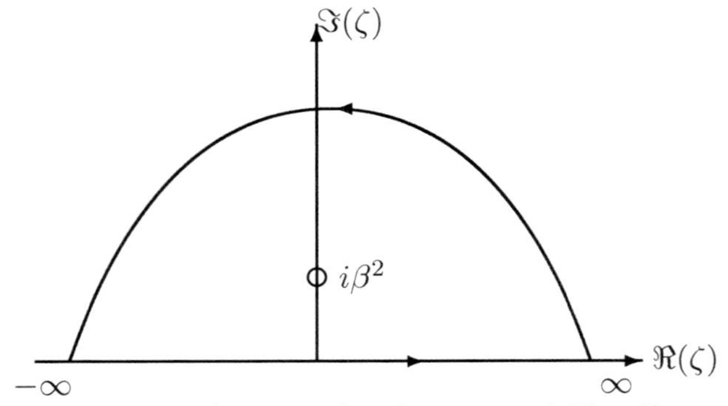

Figure 1.7: Contour for the integral $I(\alpha, \beta)$.

where Γ is a simple closed path enclosing all these isolated points. If $f(z)$ has a pole of order N at z_0, the following formula gives a quick way to calculate the residue

$$\text{Res}f(z)|_{z_0} = \frac{1}{(N-1)!} \lim_{z \to z_0} \frac{d^{N-1}[(z-z_0)^N f(z)]}{dz^{N-1}}. \quad (1.138)$$

The residue theorem serves a powerful tool for calculating some real integrals and summation of series, especially when the integrand is a function of sin and cos that can be changed into a complex integral. The real integral $\int_{-\infty}^{\infty} \psi(x)dx$ becomes $2\pi i$ multiplying the sum of the residues of $\psi(x)$ at the poles in the upper half-space.

◇ **Example 1.5:** In order to evaluate the integral

$$I(\alpha, \beta) = \int_{\infty}^{\infty} \frac{e^{i\alpha^2 \zeta}}{x^2 + \beta^4} d\zeta,$$

it is necessary to construct a contour (see Figure 1.7). As the function $\phi = e^{i\alpha^2\zeta}/(\beta^4 + \zeta^2)$ has two poles $\zeta = +i\beta^2$ and $-i\beta^2$ from $\beta^4 + \zeta^2 = 0$, and only one pole $\zeta = +i\beta^2$ is in the upper half plane, we can construct a contour to encircle the pole at $\zeta = i\beta^2$ by adding an additional arc at the infinity ($\zeta \to \infty$) on the upper half plane. Combining the arc with the horizontal line from the integral limits from $-\infty$ to ∞ along the ζ-axis, a contour is closed. Hence, we have

$$\phi = \frac{e^{i\alpha^2\zeta}/(\zeta + i\beta^2)}{\zeta - i\beta^2} = \frac{f(\zeta)}{\zeta - i\beta^2},$$

where $f(\zeta) = e^{i\alpha^2\zeta}/(\zeta + i\beta^2)$. Using the residue theorem, we have

$$I = 2\pi i[f(\zeta = i\beta^2)] = 2\pi i \frac{e^{-\alpha^2\beta^2}}{i\beta^2 + i\beta^2} = \pi \frac{e^{-\alpha^2\beta^2}}{\beta^2}.$$

In a special case when $\alpha = 0$, we have

$$\int_{-\infty}^{\infty} \frac{1}{\zeta^2 + \beta^4} d\zeta = \frac{\pi}{\beta^2}.$$

◇

With these fundamentals of preliminary mathematics, we are now ready to study a wide range of numerical methods in engineering. Before we proceed to introduce various finite element methods, however, we have to introduce the fundamental concepts of vectors, matrices, and variations of calculus in the next few chapters.

Chapter 2

Vector and Matrix Algebra

Many quantities such as force, velocity, and deformation in engineering and sciences are vectors which have both a magnitude and a direction. The manipulation of vectors is often associated with matrices. In this chapter, we will introduce the basics of vectors and vector analysis.

2.1 Vector Analysis

2.1.1 Vectors

A vector $\mathbf{x}$ is a set of ordered numbers $\mathbf{x} = (x_1, x_2, ..., x_n)$, where its components $x_1, x_2, ..., x_n$ are real numbers. All these vectors form a n-dimensional vector space $\mathcal{V}^n$. To add two vectors $\mathbf{x} = (x_1, x_2, ..., x_n)$ and $\mathbf{y} = (y_1, y_2, ..., y_n)$, we simply add their corresponding components,

$$\mathbf{z} = \mathbf{x} + \mathbf{y} = (x_1 + y_1, x_2 + y_2, ..., x_n + y_n), \qquad (2.1)$$

and the sum is also a vector. This follows the vector addition as shown in Fig 2.1

The addition of vectors has commutability ($\mathbf{u} + \mathbf{v} = \mathbf{v} + \mathbf{u}$) and associativity $[(\mathbf{a} + \mathbf{b}) + \mathbf{c} = \mathbf{a} + (\mathbf{b} + \mathbf{c})]$. Zero vector

0 is a special vector that all its components are zeros. The multiplication of a vector **x** with a scalar or constant α is carried out by the multiplication of each component,

$$\alpha \mathbf{y} = (\alpha y_1, \alpha y_2, ..., \alpha y_n). \tag{2.2}$$

Thus, $-\mathbf{y} = (-y_1, -y_2, ..., -y_n)$. In addition, $(\alpha\beta)\mathbf{y} = \alpha(\beta\mathbf{y})$ and $(\alpha + \beta)\mathbf{y} = \alpha\mathbf{y} + \beta\mathbf{y}$.

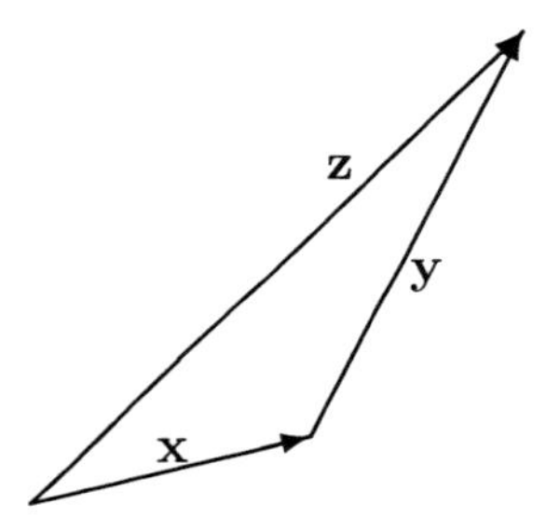

Figure 2.1: Vector addition.

Two nonzero vectors **a** and **b** are said to be linearly independent if $\alpha\mathbf{a} + \beta\mathbf{b} = 0$ implies that $\alpha = \beta = 0$. If α, β are not all zeros, then these two vectors are linearly dependent. Two linearly dependent vectors are parallel ($\mathbf{a} /\!/ \mathbf{b}$) to each other. Three linearly dependent vectors $\mathbf{a}, \mathbf{b}, \mathbf{c}$ are in the same plane.

Dot Product and Norm

The dot product or inner product of two vectors **x** and **y** is defined as

$$\mathbf{x} \cdot \mathbf{y} = x_1 y_1 + x_2 y_2 + ... + x_n y_n = \sum_{i=1}^{n} x_i y_i, \tag{2.3}$$

which is a real number. The length or norm of a vector $\mathbf{x}$ is the root of the dot product of the vector itself,

$$|\mathbf{x}| = ||\mathbf{x}|| = \sqrt{\mathbf{x} \cdot \mathbf{x}} = \sqrt{\sum_{i=1}^{n} x_i^2}. \tag{2.4}$$

When $||\mathbf{x}|| = 1$, then it is a unit vector. It is straightforward to check that the dot product has the following properties:

$$\mathbf{x} \cdot \mathbf{y} = \mathbf{y} \cdot \mathbf{x}, \qquad \mathbf{x} \cdot (\mathbf{y} + \mathbf{z}) = \mathbf{x} \cdot \mathbf{y} + \mathbf{x} \cdot \mathbf{z}, \tag{2.5}$$

and

$$(\alpha \mathbf{x}) \cdot (\beta \mathbf{y}) = (\alpha\beta) \mathbf{x} \cdot \mathbf{y}, \tag{2.6}$$

where α, β are constants.

If θ is the angle between two vectors $\mathbf{a}$ and $\mathbf{b}$, then the dot product can also be written

$$\mathbf{a} \cdot \mathbf{b} = ||\mathbf{a}|| \ ||\mathbf{b}|| \cos(\theta), \qquad 0 \le \theta \le \pi. \tag{2.7}$$

If the dot product of these two vectors is zero or $\cos(\theta) = 0$ (i.e., $\theta = \pi/2$), then we say that these two vectors are orthogonal.

Rearranging equation (2.7), we obtain a formula to calculate the angle θ between two vectors

$$\cos(\theta) = \frac{\mathbf{a} \cdot \mathbf{b}}{||\mathbf{a}|| \ ||\mathbf{b}||}. \tag{2.8}$$

Since $|\cos(\theta)| \le 1$, then we get the useful Cauchy-Schwartz inequality:

$$||\mathbf{a} \cdot \mathbf{b}|| \le ||\mathbf{a}|| \ ||\mathbf{b}||. \tag{2.9}$$

Any vector $\mathbf{a}$ in a n-dimensional vector space $\mathcal{V}^n$ can be written as a combination of a set of n independent basis vectors or orthogonal spanning vectors $\mathbf{e_1}, \mathbf{e_2}, ..., \mathbf{e_n}$, so that

$$\mathbf{a} = \alpha_1 \mathbf{e}_1 + \alpha_2 \mathbf{e}_2 + ... + \alpha_n \mathbf{e}_n = \sum_{i=1}^{n} \alpha_i \mathbf{e}_i, \tag{2.10}$$

where the coefficients/scalars $\alpha_1, \alpha_2, ..., \alpha_n$ are the components of $\mathbf{a}$ relative to the basis $\mathbf{e_1}, \mathbf{e_2} ..., \mathbf{e_n}$. The most common basis vectors are the orthogonal unit vectors. In a three-dimensional case, they are $\mathbf{i} = (1, 0, 0)$, $\mathbf{j} = (0, 1, 0$, $\mathbf{k} = (0, 0, 1)$ for three x-, y-, z-axis, and thus $\mathbf{x} = x_1\mathbf{i} + x_2\mathbf{j} + x_3\mathbf{k}$. The three unit vectors satisfy $\mathbf{i} \cdot \mathbf{j} = \mathbf{j} \cdot \mathbf{k} = \mathbf{k} \cdot \mathbf{i} = 0$.

Cross Product

The dot product of two vectors is a scalar or a number. On the other hand, the cross product or outer product of two vectors is a new vector

$$\mathbf{c} = \mathbf{a} \times \mathbf{b}$$

$$= (x_2y_3 - x_3y_2)\mathbf{i} + (x_3y_1 - x_1y_3)\mathbf{j} + (x_1y_2 - x_2y_1)\mathbf{k}, \quad (2.11)$$

which is usually written as

$$\mathbf{a} \times \mathbf{b} = \begin{vmatrix} \mathbf{i} & \mathbf{j} & \mathbf{k} \\ x_1 & x_2 & x_3 \\ y_1 & y_2 & y_3 \end{vmatrix}$$

$$= \begin{vmatrix} x_2 & x_3 \\ y_2 & y_3 \end{vmatrix} \mathbf{i} + \begin{vmatrix} x_3 & x_1 \\ y_3 & y_1 \end{vmatrix} \mathbf{j} + \begin{vmatrix} x_1 & x_2 \\ y_1 & y_2 \end{vmatrix} \mathbf{k}. \quad (2.12)$$

The angle between $\mathbf{a}$ and $\mathbf{b}$ can also be expressed as

$$\sin\theta = \frac{\|\mathbf{a} \times \mathbf{b}\|}{\|\mathbf{a}\| \, \|\mathbf{b}\|}. \quad (2.13)$$

In fact, the norm $\|\mathbf{a} \times \mathbf{b}\|$ is the area of the parallelogram formed by $\mathbf{a}$ and $\mathbf{b}$. The vector $\mathbf{c} = \mathbf{a} \times \mathbf{b}$ is perpendicular to both $\mathbf{a}$ and $\mathbf{b}$, following a right-hand rule. It is straightforward to check that the cross product has the following properties:

$$\mathbf{x}\times\mathbf{y} = -\mathbf{y}\times\mathbf{x}, \quad (\mathbf{x} + \mathbf{y})\times\mathbf{z} = \mathbf{x}\times\mathbf{z} + \mathbf{y}\times\mathbf{z}, \quad (2.14)$$

and

$$(\alpha\mathbf{x})\times(\beta\mathbf{y}) = (\alpha\beta)\mathbf{x}\times\mathbf{y}. \quad (2.15)$$

A very special case is $\mathbf{a}\times\mathbf{a} = \mathbf{0}$. For unit vectors, we have

$$\mathbf{i}\times\mathbf{j} = \mathbf{k}, \qquad \mathbf{j}\times\mathbf{k} = \mathbf{i}, \qquad \mathbf{k}\times\mathbf{i} = \mathbf{j}. \quad (2.16)$$

◇ **Example 2.1:** For two 3-D vectors $\mathbf{a} = (1, 1, 0)$ and $\mathbf{b} = (2, -1, 0)$, their dot product is

$$\mathbf{a} \cdot \mathbf{b} = 1 \times 2 + 1 \times (-1) + 0 = 1.$$

As their moduli are

$$||\mathbf{a}|| = \sqrt{1^2 + 1^2 + 0^2} = \sqrt{2}, \quad ||\mathbf{b}|| = \sqrt{2^2 + (-1)^2 + 0} = \sqrt{5},$$

we can calculate the angle θ between the two vectors. We have

$$\cos\theta = \frac{\mathbf{a} \cdot \mathbf{b}}{||\mathbf{a}|| ||\mathbf{b}||} = \frac{1}{\sqrt{2}\sqrt{5}},$$

or

$$\theta = \cos^{-1} \frac{1}{\sqrt{10}} \approx 71.56^\circ.$$

Their cross product is

$$\mathbf{v} = \mathbf{a} \times \mathbf{b} = (1 \times 0 - 0 \times (-1), 0 \times 1 - 1 \times 0, 1 \times (-1) - 2 \times 1)$$

$$= (0, 0, -3),$$

which is a vector pointing in the negative z-axis direction. The vector $\mathbf{v}$ is perpendicular to both $\mathbf{a}$ and $\mathbf{b}$ because

$$\mathbf{a} \cdot \mathbf{v} = 1 \times 0 + 1 \times 0 + 0 \times (-3) = 0,$$

and

$$\mathbf{b} \cdot \mathbf{v} = 2 \times 0 + (-1) \times 0 + 0 \times (-3) = 0.$$

--- ◇

Vector Triple

For two vectors, their product can be either a scalar (dot product) or a vector (cross product). Similarly, the product of triple vectors $\mathbf{a}, \mathbf{b}, \mathbf{c}$ can be either a scalar

$$\mathbf{a} \cdot (\mathbf{b} \times \mathbf{c}) = \begin{vmatrix} a_x & a_y & a_z \\ b_x & b_y & b_z \\ c_x & c_y & c_z \end{vmatrix}, \tag{2.17}$$

or a vector

$$\mathbf{a} \times (\mathbf{b} \times \mathbf{c}) = (\mathbf{a} \cdot \mathbf{c})\mathbf{b} - (\mathbf{a} \cdot \mathbf{b})\mathbf{c}. \tag{2.18}$$

As the dot product of two vectors is the area of a parallelogram, the scalar triple product is the volume of the parallelepiped formed by the three vectors. From the definitions, it is straightforward to prove that

$$\mathbf{a} \cdot (\mathbf{b} \times \mathbf{c}) = \mathbf{b} \cdot (\mathbf{c} \times \mathbf{a}) = \mathbf{c} \cdot (\mathbf{a} \times \mathbf{b}) = -\mathbf{a} \cdot (\mathbf{c} \times \mathbf{b}), \tag{2.19}$$

$$\mathbf{a} \times (\mathbf{b} \times \mathbf{c}) \neq (\mathbf{a} \times \mathbf{b}) \times \mathbf{c}, \tag{2.20}$$

and

$$(\mathbf{a} \times \mathbf{b}) \cdot (\mathbf{c} \times \mathbf{d}) = (\mathbf{a} \cdot \mathbf{c})(\mathbf{b} \cdot \mathbf{d}) - (\mathbf{a} \cdot \mathbf{d})(\mathbf{b} \cdot \mathbf{c}). \tag{2.21}$$

2.1.2 Vector Calculus

Differentiation of Vectors

The differentiation of a vector is carried out over each component and treating each component as the usual differentiation of a scalar. Thus, for a position vector

$$\mathbf{P}(t) = x(t)\mathbf{i} + y(t)\mathbf{j} + z(t)\mathbf{k}, \tag{2.22}$$

we can write its velocity as

$$\mathbf{v} = \frac{d\mathbf{P}}{dt} = \dot{x}(t)\mathbf{i} + \dot{y}(t)\mathbf{j} + \dot{z}(t)\mathbf{k}, \tag{2.23}$$

and acceleration as

$$\mathbf{a} = \frac{d^2\mathbf{P}}{dt^2} = \ddot{x}(t)\mathbf{i} + \ddot{y}(t)\mathbf{j} + \ddot{z}(t)\mathbf{k}, \tag{2.24}$$

where $\dot{()} = d()/dt$. Conversely, the integral of $\mathbf{v}$ is

$$\mathbf{P} = \int \mathbf{v} dt + \mathbf{c}, \tag{2.25}$$

where $\mathbf{c}$ is a vector constant.

From the basic definition of differentiation, it is easy to check that the differentiation of vectors has the following properties:

$$\frac{d(\alpha \mathbf{a})}{dt} = \alpha \frac{d\mathbf{a}}{dt}, \quad \frac{d(\mathbf{a} \cdot \mathbf{b})}{dt} = \frac{d\mathbf{a}}{dt} \cdot \mathbf{b} + \mathbf{a} \cdot \frac{d\mathbf{b}}{dt}, \tag{2.26}$$

and

$$\frac{d(\mathbf{a} \times \mathbf{b})}{dt} = \frac{d\mathbf{a}}{dt} \times \mathbf{b} + \mathbf{a} \times \frac{d\mathbf{b}}{dt}. \tag{2.27}$$

Line Integral

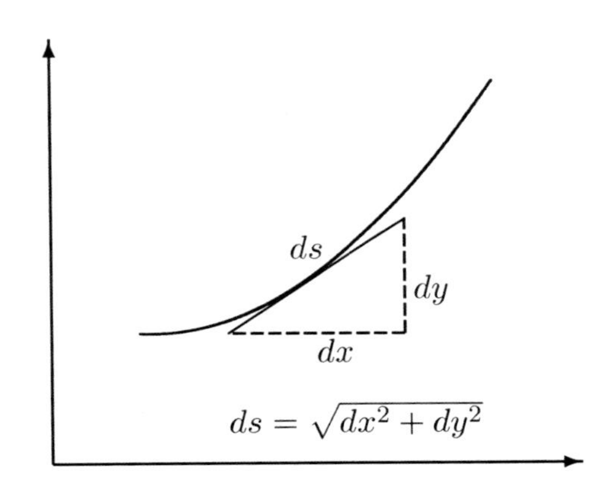

Figure 2.2: Arc length along a curve.

An important class of integrals in this context is the line integral which integrates along a curve $\mathbf{r}(x, y, z) = x\mathbf{i} + y\mathbf{j} + z\mathbf{k}$. For example, in order to calculate the arc length L of curve $\mathbf{r}$ as shown in Figure 2.2, we have to use the line integral.

$$L = \int_{s_0}^{s} ds = \int_{s_0}^{s} \sqrt{dx^2 + dy^2} = \int_{x_0}^{x} \sqrt{1 + (\frac{dy}{dx})^2} dx. \tag{2.28}$$

◇ **Example 2.2:** The arc length of the parabola $y(x) = \frac{1}{2}x^2$ from $x = 0$ to $x = 1$ is given by

$$L = \int_0^1 \sqrt{1 + y'^2} dx = \int_0^1 \sqrt{1 + x^2} dx$$

$$= \frac{1}{2}[x\sqrt{1+x^2} + \ln(x + \sqrt{1+x^2})]\Big|_0^1$$

$$= \frac{1}{2}[\sqrt{2} - \ln(\sqrt{2} - 1)] \approx 1.14779.$$

◇

Three Basic Operators

Three important operators commonly used in vector analysis, especially in fluid dynamics, are the gradient operator (grad or ∇), the divergence operator (div or $\nabla\cdot$) and the curl operator (curl or $\nabla\times$).

Sometimes, it is useful to calculate the directional derivative of a function ϕ at the point (x, y, z) in the direction of $\mathbf{n}$

$$\frac{\partial \phi}{\partial \mathbf{n}} = \mathbf{n} \cdot \nabla \phi = \frac{\partial \phi}{\partial x}\cos(\alpha) + \frac{\partial \phi}{\partial y}\cos(\beta) + \frac{\partial \phi}{\partial z}\cos(\gamma), \quad (2.29)$$

where $\mathbf{n} = (\cos\alpha, \cos\beta, \cos\gamma)$ is a unit vector and α, β, γ are the directional angles. Generally speaking, the gradient of any scalar function ϕ of x, y, z can be written in a similar way,

$$\text{grad}\phi = \nabla\phi = \frac{\partial \phi}{\partial x}\mathbf{i} + \frac{\partial \phi}{\partial y}\mathbf{j} + \frac{\partial \phi}{\partial z}\mathbf{k}. \quad (2.30)$$

This is equivalent to applying the del operator ∇ to the scalar function ϕ

$$\nabla = \frac{\partial}{\partial x}\mathbf{i} + \frac{\partial}{\partial y}\mathbf{j} + \frac{\partial}{\partial z}\mathbf{k}. \quad (2.31)$$

The direction of the gradient operator on a scalar field gives a vector field. The gradient operator has the following properties:

$$\nabla(\alpha\psi + \beta\phi) = \alpha\nabla\psi + \beta\nabla\phi, \qquad \nabla(\psi\phi) = \psi\nabla\phi + \phi\nabla\psi, \quad (2.32)$$

where α, β are constants and ψ, ϕ are scalar functions.

For a vector field

$$\mathbf{u}(x, y, z) = u_1(x, y, z)\mathbf{i} + u_2(x, y, z)\mathbf{j} + u_3(x, y, z)\mathbf{k}, \quad (2.33)$$

the application of the operator ∇ can lead to either a scalar field or a vector field, depending on how the del operator applies to the vector field. The divergence of a vector field is the dot product of the del operator ∇ and $\mathbf{u}$

$$\text{div } \mathbf{u} = \nabla \cdot \mathbf{u} = \frac{\partial \mathrm{u}_1}{\partial \mathrm{x}} + \frac{\partial \mathrm{u}_2}{\partial \mathrm{y}} + \frac{\partial \mathrm{u}_3}{\partial \mathrm{z}}, \tag{2.34}$$

and the curl of $\mathbf{u}$ is the cross product of the del operator and the vector field $\mathbf{u}$

$$\text{curl } \mathbf{u} = \nabla \times \mathbf{u} = \begin{vmatrix} \mathbf{i} & \mathbf{j} & \mathbf{k} \\ \frac{\partial}{\partial x} & \frac{\partial}{\partial y} & \frac{\partial}{\partial z} \\ u_1 & u_2 & u_3 \end{vmatrix}. \tag{2.35}$$

It is straightforward to verify the following useful identities associated with the ∇ operator:

$$\nabla \cdot \nabla \times \mathbf{u} = 0, \tag{2.36}$$

$$\nabla \times \nabla \psi = 0, \tag{2.37}$$

$$\nabla \times (\psi \mathbf{u}) = \psi \nabla \times \mathbf{u} + (\nabla \psi) \times \mathbf{u}, \tag{2.38}$$

$$\nabla \cdot (\psi \mathbf{u}) = \psi \nabla \cdot \mathbf{u} + (\nabla \psi) \cdot \mathbf{u}, \tag{2.39}$$

$$\nabla \times (\nabla \times \mathbf{u}) = \nabla(\nabla \cdot \mathbf{u}) - \nabla^{\mathbf{2}} \mathbf{u}. \tag{2.40}$$

One of the most common operators in engineering and science is the Laplacian operator is

$$\nabla^2 \Psi = \nabla \cdot (\nabla \Psi) = \frac{\partial^2 \Psi}{\partial x^2} + \frac{\partial^2 \Psi}{\partial y^2} + \frac{\partial^2 \Psi}{\partial z^2}, \tag{2.41}$$

for Laplace's equation

$$\Delta \Psi = \nabla^2 \Psi = 0. \tag{2.42}$$

It is sometimes necessary to express the Laplace equation in other coordinates. In cylindrical polar coordinates (r, ϕ, z), we have

$$\nabla \cdot \mathbf{u} = \frac{1}{r} \frac{\partial (r u_r)}{\partial r} + \frac{1}{r} \frac{\partial u_\phi}{\partial \phi} + \frac{\partial u_z}{\partial z}. \tag{2.43}$$

The Laplace equation becomes

$$\nabla^2 \Psi = \frac{\partial^2 \Psi}{\partial r^2} + \frac{1}{r}\frac{\partial \Psi}{\partial r} + \frac{1}{r^2}\frac{\partial^2 \Psi}{\partial \phi^2} + \frac{\partial^2 \Psi}{\partial z^2}. \tag{2.44}$$

In spherical polar coordinates (r, θ, ϕ), we have

$$\nabla \cdot \mathbf{u} = \frac{1}{r^2}\frac{\partial^2 (r^2 u_r)}{\partial r^2} + \frac{1}{r \sin\theta}\frac{\partial(\sin\theta u_\theta)}{\partial \theta} + \frac{1}{r \sin\theta}\frac{\partial u_\phi}{\partial \phi}. \tag{2.45}$$

The Laplace equation can be written as

$$\nabla^2 \Psi = \frac{1}{r^2}\frac{\partial}{\partial r}[r^2 \frac{\partial \Psi}{\partial r}]$$

$$+\frac{1}{r^2 \sin\theta}\frac{\partial}{\partial \theta}[\sin\theta \frac{\partial \Psi}{\partial \theta}] + \frac{1}{r^2 \sin^2\theta}\frac{\partial^2 \Psi}{\partial \phi^2}. \tag{2.46}$$

Some Important Theorems

The Green theorem is an important theorem, especially in fluid dynamics and the finite element analysis. For a vector field $\mathbf{Q} = u\mathbf{i} + v\mathbf{j}$ in a 2-D region Ω with the boundary Γ and the unit outer normal $\mathbf{n}$ and unit tangent $\mathbf{t}$. The theorems connect the integrals of divergence and curl with other integrals. Gauss's theorem states:

$$\iiint_\Omega (\nabla \cdot \mathbf{Q}) d\Omega = \iint_S \mathbf{Q} \cdot \mathbf{n} dS, \tag{2.47}$$

which connects the volume integral to the surface integral.

Another important theorem is Stokes's theorem:

$$\iint_S (\nabla \times \mathbf{Q}) \cdot \mathbf{k} d\mathbf{S} = \oint_\Gamma \mathbf{Q} \cdot \mathbf{t} d\Gamma = \oint_\Gamma \mathbf{Q} \cdot d\mathbf{r}, \tag{2.48}$$

which connects the surface integral to the corresponding line integral.

In our simple 2-D case, this becomes

$$\oint (u dx + v dy) = \iint_\Omega (\frac{\partial v}{\partial x} - \frac{\partial u}{\partial y}) dx dy. \tag{2.49}$$

For any scalar functions ψ and ϕ, the useful Green's first identity can be written as

$$\oint_{\partial\Omega} \psi\nabla\phi d\Gamma = \int_{\Omega} (\psi\nabla^2\phi + \nabla\psi \cdot \nabla\phi) d\Omega, \tag{2.50}$$

where $d\Omega = dxdydz$. By using this identity twice, we get Green's second identity

$$\oint_{\partial\Omega} (\psi\nabla\phi - \phi\nabla\psi) d\Gamma = \int_{\Omega} (\psi\nabla^2\phi - \phi\nabla^2\psi) d\Omega. \tag{2.51}$$

2.1.3 Conservation of Mass

The mass conservation in flow mechanics can be expressed in either integral form (weak form) or differential form (strong form). For any enclosed volume Ω, the total mass which leaves or enters the surface S is

$$\oint_S \rho \boldsymbol{u} \cdot d\mathbf{A},$$

where $\rho(x, y, z, t)$ and $\boldsymbol{u}(x, y, z, t)$ are the density and the velocity of the fluid, respectively. The rate of change of mass in Ω is

$$\frac{\partial}{\partial t} \int \rho dV.$$

The mass conservation requires that the rate of loss of mass through the surface S is balanced by the rate of change in Ω. Therefore, we have

$$\oint_S \rho \boldsymbol{u} \cdot d\mathbf{A} + \frac{\partial}{\partial t} \int dV = 0.$$

Using Gauss's theorem for the surface integral, we have

$$\int_{\Omega} \nabla \cdot (\rho \boldsymbol{u}) dV + \frac{\partial}{\partial t} \int_{\Omega} \rho dV = 0.$$

Interchange of the integration and differentiation in the second term, we have

$$\int_{\Omega} [\frac{\partial \rho}{\partial t} + \nabla \cdot (\rho \boldsymbol{u})] dV = 0.$$

This is the integral form or weak form of the conservation of mass. This is true for any volume at any instance, and subsequently the only way that this is true for all possible choice of Ω is

$$\frac{\partial \rho}{\partial t} + \nabla \cdot (\rho \boldsymbol{u}) = 0,$$

which is the differential form or strong form of mass conservation. Other laws of conservation such as conservation of momentum and energy can be derived in a similar manner. The integral form is more useful in numerical methods such as finite volume methods and finite element methods, while the differential form is more natural for mathematical analysis.

2.2 Matrix Algebra

2.2.1 Matrix

Matrices are widely used in engineering and computational sciences. A matrix is a table or array of numbers or functions arranged in rows and columns. The elements or entries of a matrix $\mathbf{A}$ are often denoted as a_{ij}. A matrix $\mathbf{A}$ has m rows and n columns,

$$\mathbf{A} = [a_{ij}] = \begin{pmatrix} a_{11} & a_{12} & \ldots & a_{1j} & \ldots & a_{1n} \\ a_{21} & a_{22} & \ldots & a_{2j} & \ldots & a_{2n} \\ \vdots & \vdots & & a_{ij} & \ldots & \vdots \\ a_{m1} & a_{m2} & \ldots & a_{mj} & \ldots & a_{mn} \end{pmatrix}, \tag{2.52}$$

we say the size of $\mathbf{A}$ is m by n, or $m \times n$. $\mathbf{A}$ is square if $m = n$. For example,

$$\mathbf{A} = \begin{pmatrix} 1 & 2 & 3 \\ 4 & 5 & 6 \end{pmatrix}, \qquad \mathbf{B} = \begin{pmatrix} e^x & \sin x \\ -i\cos x & e^{i\theta} \end{pmatrix}, \tag{2.53}$$

and

$$\mathbf{u} = \begin{pmatrix} u \\ v \\ w \end{pmatrix}, \tag{2.54}$$

where $\mathbf{A}$ is a 2×3 matrix, $\mathbf{B}$ is a 2×2 square matrix, and $\mathbf{u}$ is a 3×1 column matrix or column vector.

The sum of two matrices $\mathbf{A}$ and $\mathbf{B}$ is only possible if they have the same size $m\times n$, and their sum, which is also $m\times n$, is obtained by adding corresponding entries

$$\mathbf{C} = \mathbf{A} + \mathbf{B}, \qquad c_{ij} = a_{ij} + b_{ij}, \tag{2.55}$$

where $(i = 1, 2, ..., m; j = 1, 2, ..., n)$. We can multiply a matrix $\mathbf{A}$ by a scalar α by multiplying each entry by α. The product of two matrices is only possible if the number of columns of $\mathbf{A}$ is the same as the number of rows of $\mathbf{B}$. That is to say, if $\mathbf{A}$ is $m\times n$ and $\mathbf{B}$ is $n\times r$, then the product $\mathbf{C}$ is $m\times r$,

$$c_{ij} = (AB)_{ij} = \sum_{k=1}^{n} a_{ik} b_{kj}. \tag{2.56}$$

If $\mathbf{A}$ is a square matrix, then we have $\mathbf{A}^n = \overbrace{AA...A}^{n}$. The multiplications of matrices are generally not commutive, i.e., $\mathbf{AB} \neq \mathbf{BA}$. However, the multiplication has associativity $\mathbf{A}(\mathbf{uv}) = (\mathbf{Au})\mathbf{v}$ and $\mathbf{A}(\mathbf{u} + \mathbf{v}) = \mathbf{Au} + \mathbf{Av}$.

The transpose $\mathbf{A}^T$ of $\mathbf{A}$ is obtained by switching the position of rows and columns, and thus $\mathbf{A}^T$ will be $n\times m$ if $\mathbf{A}$ is $m\times n$, $(a^T)_{ij} = a_{ji}, (i = 1, 2, ..., m; j = 1, 2, ..., n)$. In general, we have

$$(\mathbf{A}^T)^T = \mathbf{A}, \qquad (\mathbf{AB})^T = \mathbf{B}^T\mathbf{A}^T. \tag{2.57}$$

The differentiation and integral of a matrix are done on each member element. For example, for a 2×2 matrix

$$\frac{d\mathbf{A}}{dt} = \dot{\mathbf{A}} = \begin{pmatrix} \frac{da_{11}}{dt} & \frac{da_{12}}{dt} \\ \frac{da_{21}}{dt} & \frac{da_{22}}{dt} \end{pmatrix}, \tag{2.58}$$

and

$$\int \mathbf{A}dt = \begin{pmatrix} \int a_{11}dt & \int a_{12}dt \\ \int a_{21}dt & \int a_{22}dt \end{pmatrix}. \tag{2.59}$$

A diagonal matrix $\mathbf{A}$ is a square matrix whose every entry off the main diagonal is zero ($a_{ij} = 0$ if $i \neq j$). Its diagonal elements or entries may or may not have zeros. For example, the matrix

$$\mathbf{I} = \begin{pmatrix} 1 & 0 & 0 \\ 0 & 1 & 0 \\ 0 & 0 & 1 \end{pmatrix} \tag{2.60}$$

is a 3×3 identity or unitary matrix. In general, we have

$$\mathbf{AI} = \mathbf{IA} = \mathbf{A}. \tag{2.61}$$

A zero or null matrix $\mathbf{0}$ is a matrix with all of its elements being zero.

Determinant

The determinant of a square matrix $\mathbf{A}$ is a number or scalar obtained by the following recursive formula or the cofactor or Laplace expansion by column or row. For example, expanding by row k, we have

$$\det(\mathbf{A}) = |\mathbf{A}| = \sum_{j=1}^{n} (-1)^{k+j} a_{kj} M_{kj}, \tag{2.62}$$

where M_{ij} is the determinant of a minor matrix of $\mathbf{A}$ obtained by deleting row i and column j. For a simple 2×2 matrix, its determinant simply becomes

$$\begin{vmatrix} a_{11} & a_{12} \\ a_{21} & a_{22} \end{vmatrix} = a_{11}a_{22} - a_{12}a_{21}. \tag{2.63}$$

It is easy to verify that the determinant has the following properties:

$$|\alpha\mathbf{A}| = \alpha|\mathbf{A}|, \qquad |\mathbf{A}^T| = |\mathbf{A}|, \qquad |\mathbf{AB}| = |\mathbf{A}||\mathbf{B}|, \tag{2.64}$$

where $\mathbf{A}$ and $\mathbf{B}$ are the same size ($n \times n$).

A $n \times n$ square matrix is singular if $|\mathbf{A}| = 0$, and is nonsingular if and only if $|\mathbf{A}| \neq 0$. The trace of a square matrix $\mathrm{tr}(\mathbf{A})$ is defined as the sum of the diagonal elements,

$$\mathrm{tr}(\mathbf{A}) = \sum_{\mathrm{i}=1}^{\mathrm{n}} a_{ii} = a_{11} + a_{22} + ... + a_{nn}. \tag{2.65}$$

The rank of a matrix $\mathbf{A}$ is the number of linearly independent vectors forming the matrix. Generally, the rank of $\mathbf{A}$ is $\text{rank}(\mathbf{A}) \leq \min(\text{m}, \text{n})$. For a $n \times n$ square matrix $\mathbf{A}$, it is nonsingular if $\text{rank}(\mathbf{A}) = n$.

From the basic definitions, it is straightforward to prove the following

$$(\mathbf{AB...Z})^{\mathbf{T}} = \mathbf{Z}^{\mathbf{T}}...\mathbf{B}^{\mathbf{T}}\mathbf{A}^{\mathbf{T}}, \tag{2.66}$$

$$|\mathbf{AB....Z}| = |\mathbf{A}||\mathbf{B}|...|\mathbf{Z}|, \tag{2.67}$$

$$\text{tr}(\mathbf{A}) = \text{tr}(\mathbf{A}^T), \;\; \text{tr}(\mathbf{A}+\mathbf{B}) = \text{tr}(\mathbf{A}) + \text{tr}(\mathbf{B}), \tag{2.68}$$

$$\text{tr}(\mathbf{AB}) = \text{tr}(\mathbf{BA}), \;\; \det(\mathbf{AB}) = \det(\mathbf{A})\det(\mathbf{B}). \tag{2.69}$$

Inverse

The inverse matrix $\mathbf{A}^{-1}$ of a square matrix $\mathbf{A}$ is defined as

$$\mathbf{A}^{-1}\mathbf{A} = \mathbf{A}\mathbf{A}^{-1} = \mathbf{I}. \tag{2.70}$$

It is worth noting that the unit matrix $\mathbf{I}$ has the same size as $\mathbf{A}$. The inverse of a square matrix exists if and only if $\mathbf{A}$ is nonsingular or $\det(\mathbf{A}) \neq 0$. From the basic definitions, it is straightforward to prove that the inverse of a matrix has the following properties

$$(\mathbf{A}^{-1})^{-1} = \mathbf{A}, \qquad (\mathbf{A}^{\mathbf{T}})^{-1} = (\mathbf{A}^{-1})^{\mathbf{T}}, \tag{2.71}$$

and

$$(\mathbf{AB})^{-1} = \mathbf{B}^{-1}\mathbf{A}^{-1}. \tag{2.72}$$

A simple useful formula for obtaining the inverse of a 2×2 matrix is

$$\begin{pmatrix} a & b \\ c & d \end{pmatrix}^{-1} = \frac{1}{(ad-bc)} \begin{pmatrix} d & -b \\ -c & a \end{pmatrix}. \tag{2.73}$$

◇ **Example 2.3:** For two matrices

$$\mathbf{A} = \begin{pmatrix} 1 & 2 & 3 \\ -1 & 1 & 0 \\ 3 & 2 & 2 \end{pmatrix}, \quad \mathbf{B} = \begin{pmatrix} 1 & -1 \\ 2 & 3 \\ 1 & 7 \end{pmatrix},$$

we have

$$\mathbf{AB} = \mathbf{V} = \begin{pmatrix} V_{11} & V_{12} \\ V_{21} & V_{22} \\ V_{31} & V_{32} \end{pmatrix},$$

where

$$V_{11} = 1 \times 1 + 2 \times 2 + 3 \times 1 = 8, \; V_{12} = 1 \times (-1) + 2 \times 3 + 3 \times 7 = 26;$$

$$V_{21} = -1 \times 1 + 1 \times 2 + 0 \times 1 = 1, \; V_{22} = -1 \times (-1) + 1 \times 3 + 0 \times 7 = 4;$$

$$V_{31} = 3 \times 1 + 2 \times 2 + 2 \times 1 = 9, \; V_{32} = 3 \times (-1) + 2 \times 3 + 2 \times 7 = 17.$$

Thus,

$$\mathbf{AB} = \mathbf{V} = \begin{pmatrix} 8 & 26 \\ 1 & 4 \\ 9 & 17 \end{pmatrix}.$$

However, $\mathbf{BA}$ does not exist. The transpose matrices of $\mathbf{A}$ and $\mathbf{B}$ are

$$\mathbf{A}^T = \begin{pmatrix} 1 & -1 & 3 \\ 2 & 1 & 2 \\ 3 & 0 & 2 \end{pmatrix}, \qquad \mathbf{B}^T = \begin{pmatrix} 1 & 2 & 1 \\ -1 & 3 & 7 \end{pmatrix}.$$

Similarly, we have

$$\mathbf{B}^T \mathbf{A}^T = \begin{pmatrix} 8 & 1 & 9 \\ 26 & 4 & 17 \end{pmatrix} = \mathbf{V}^T = (\mathbf{AB})^T.$$

The inverse of $\mathbf{A}$ is

$$\mathbf{A}^{-1} = \frac{1}{9} \begin{pmatrix} -2 & -2 & 3 \\ -2 & 7 & 3 \\ 5 & -4 & -3 \end{pmatrix},$$

and the determinant of $\mathbf{A}$ is

$$\det |\mathbf{A}| = -9.$$

The trace of $\mathbf{A}$ is

$$\mathrm{tr}(\mathbf{A}) = A_{11} + A_{22} + A_{33} = 1 + 1 + 2 = 4.$$

◇

Matrix Exponential

Sometimes, we need to calculate exp[**A**], where **A** is a square matrix. In this case, we have to deal with matrix exponentials. The exponential of a square matrix **A** is defined as

$$e^{\mathbf{A}} \equiv \sum_{n=0}^{\infty} \frac{1}{n!} \mathbf{A}^n = \mathbf{I} + \mathbf{A} + \frac{1}{2}\mathbf{A}^2 + ... \tag{2.74}$$

where **I** is a unity matrix with the same size as **A**, and $\mathbf{A}^2 = \mathbf{AA}$ and so on. This (rather odd) definition in fact provides a method to calculate the matrix exponential. The matrix exponentials are very useful in solving systems of differential equations.

For a simple matrix

$$\mathbf{A} = \begin{pmatrix} t & 0 \\ 0 & t \end{pmatrix}, \tag{2.75}$$

we have

$$e^{\mathbf{A}} = \begin{pmatrix} e^t & 0 \\ 0 & e^t \end{pmatrix}. \tag{2.76}$$

For

$$\mathbf{A} = \begin{pmatrix} t & t \\ t & t \end{pmatrix}, \tag{2.77}$$

we have

$$e^{\mathbf{A}} = \begin{pmatrix} \frac{1}{2}(1+e^{2t}) & \frac{1}{2}(e^{2t}-1) \\ \frac{1}{2}(e^{2t}-1) & \frac{1}{2}(1+e^{2t}) \end{pmatrix}. \tag{2.78}$$

As you see, it is quite complicated but still straightforward to calculate the matrix exponentials. Fortunately, it can be easily done using a computer. By using the power expansions and the basic definition, we can prove the following useful identities

$$e^{t\mathbf{A}} \equiv \sum_{n=0}^{\infty} \frac{1}{n!}(t\mathbf{A})^n = \mathbf{I} + t\mathbf{A} + \frac{t^2}{2}\mathbf{A}^2 + ..., \tag{2.79}$$

$$\ln(\mathbf{IA}) \equiv \sum_{n=1}^{\infty} \frac{(-1)^{n-1}}{n!}\mathbf{A}^n = \mathbf{A} - \frac{1}{2}\mathbf{A}^2 + \frac{1}{3}\mathbf{A}^3 + ..., \tag{2.80}$$

$$e^{\mathbf{A}} e^{\mathbf{B}} = e^{\mathbf{A}+\mathbf{B}} \qquad (\text{if } \mathbf{AB} = \mathbf{BA}), \tag{2.81}$$

$$\frac{d}{dt} e^{t\mathbf{A}} = \mathbf{A} e^{t\mathbf{A}} = e^{t\mathbf{A}} \mathbf{A}, \tag{2.82}$$

$$(e^{\mathbf{A}})^{-1} = e^{-\mathbf{A}}, \qquad \det(e^{\mathbf{A}}) = e^{\mathrm{tr}\mathbf{A}}. \tag{2.83}$$

Hermitian and Quadratic Forms

The matrices we have discussed so far are real matrices because all their elements are real. In general, the entries or elements of a matrix can be complex numbers, and the matrix becomes a complex matrix. For a matrix $\mathbf{A}$, its complex conjugate $\mathbf{A}^*$ is obtained by taking the complex conjugate of each of its elements. The Hermitian conjugate $\mathbf{A}^\dagger$ is obtained by taking the transpose of its complex conjugate matrix. That is to say, for

$$\mathbf{A} = \begin{pmatrix} a_{11}, & a_{12}, & \dots \\ a_{21} & a_{21} & \dots \\ \dots & \dots & \dots \end{pmatrix}, \tag{2.84}$$

we have

$$\mathbf{A}^* = \begin{pmatrix} a_{11}^* & a_{12}^* & \dots \\ a_{21}^* & a_{22}^* & \dots \\ \dots & \dots & \dots \end{pmatrix}, \tag{2.85}$$

and

$$\mathbf{A}^\dagger = (\mathbf{A}^*)^T = (\mathbf{A}^T)^* = \begin{pmatrix} a_{11}^* & a_{21}^* & \dots \\ a_{12}^* & a_{22} & \dots \\ \dots & \dots & \dots \end{pmatrix}. \tag{2.86}$$

A square matrix $\mathbf{A}$ is called orthogonal if and only if $\mathbf{A}^{-1} = \mathbf{A}^T$. If a square matrix $\mathbf{A}$ satisfies $\mathbf{A}^* = \mathbf{A}$, it is said to be an Hermitian matrix. It is an anti-Hermitian matrix if $\mathbf{A}^* = -\mathbf{A}$. If the Hermitian matrix of a square matrix $\mathbf{A}$ is equal to the inverse of the matrix (or $\mathbf{A}^\dagger = \mathbf{A}^{-1}$), it is called a unitary matrix.

◇ **Example 2.4:** For a matrix

$$\mathbf{B} = \begin{pmatrix} 2+i & 3-2i & 1 \\ e^{-i\pi} & 0 & 1-i\pi \end{pmatrix},$$

its complex conjugate $\mathbf{B}^*$ and Hermitian conjugate $\mathbf{B}^\dagger$ are

$$\mathbf{B}^* = \begin{pmatrix} 2-i & 3+2i & 1 \\ e^{i\pi} & 0 & 1+i\pi \end{pmatrix},$$

$$\mathbf{B}^\dagger = \begin{pmatrix} 2-i & e^{i\pi} \\ 3+2i & 0 \\ 1 & 1+i\pi \end{pmatrix} = (\mathbf{B}^*)^T.$$

For the rotation matrix

$$\mathbf{A} = \begin{pmatrix} \cos\theta & \sin\theta \\ -\sin\theta & \cos\theta \end{pmatrix},$$

its inverse and transpose are

$$\mathbf{A}^{-1} = \frac{1}{\cos^2\theta + \sin^2\theta} \begin{pmatrix} \cos\theta & -\sin\theta \\ \sin\theta & \cos\theta \end{pmatrix},$$

and

$$\mathbf{A}^T = \begin{pmatrix} \cos\theta & -\sin\theta \\ \sin\theta & \cos\theta \end{pmatrix}.$$

Since $\cos^2\theta + \sin^2\theta = 1$, we have $\mathbf{A}^T = \mathbf{A}^{-1}$. Therefore, the original matrix $\mathbf{A}$ is orthogonal. ◇

A very useful concept in applied mathematics and computing is quadratic forms. For a real vector $\mathbf{q}^T = (q_1, q_2, q_3, ..., q_n)$ and a real square matrix $\mathbf{A}$, a quadratic form $\psi(\mathbf{q})$ is a scalar function defined by

$$\psi(\mathbf{q}) = \mathbf{q}^T \mathbf{A} \mathbf{q}$$

$$= \begin{pmatrix} q_1 & q_2 & ... & q_n \end{pmatrix} \begin{pmatrix} A_{11} & A_{12} & ... & A_{1n} \\ A_{21} & A_{22} & ... & A_{2n} \\ ... & ... & ... & ... \\ A_{n1} & A_{n2} & ... & A_{nn} \end{pmatrix} \begin{pmatrix} q_1 \\ q_2 \\ \vdots \\ q_n \end{pmatrix}, \tag{2.87}$$

which can be written as

$$\psi(\mathbf{q}) = \sum_{i=1}^{n} \sum_{j=1}^{n} q_i A_{ij} q_j. \tag{2.88}$$

Since ψ is a scalar, it should be independent of the coordinates. In the case of a square matrix $\mathbf{A}$, ψ might be more easily evaluated in certain intrinsic coordinates $Q_1, Q_2, ..., Q_n$. An important result concerning the quadratic form is that it can always be written through appropriate transformations as

$$\psi(\mathbf{q}) = \sum_{i=1}^{n} \lambda_i Q_i^2 = \lambda_1 Q_1^2 + \lambda_2 Q_2^2 + ... \lambda_n Q_n^2. \tag{2.89}$$

The natural extension of quadratic forms is the Hermitian form that is the quadratic form for complex Hermitian matrix $\mathbf{A}$. Furthermore, the matrix $\mathbf{A}$ can be linear operators and functionals in addition to numbers.

For a vector $\mathbf{q} = (q_1, q_2)$ and the square matrix

$$\mathbf{A} = \begin{pmatrix} 1 & -2 \\ -2 & 1 \end{pmatrix}, \tag{2.90}$$

we have a quadratic form

$$\psi(\mathbf{q}) = \begin{pmatrix} q_1 & q_2 \end{pmatrix} \begin{pmatrix} 1 & -2 \\ -2 & 1 \end{pmatrix} \begin{pmatrix} q_1 \\ q_2 \end{pmatrix}$$

$$= q_1^2 - 4q_1 q_2 + q_2^2. \tag{2.91}$$

2.2.2 Linear systems

A linear system of m equations for n unknowns

$$a_{11}u_1 + a_{12}u_2 + ... + a_{1n}u_n = b_1,$$

$$a_{21}u_1 + a_{22}u_2 + ... + a_{2n}u_n = b_2,$$

$$\vdots \qquad \vdots$$

$$a_{m1}u_1 + a_{m2}u_2 + ... + a_{mn}u_n = b_n, \tag{2.92}$$

can be written in the compact form as

$$\begin{pmatrix} a_{11} & a_{12} & \dots & a_{1n} \\ a_{21} & a_{22} & \dots & a_{2n} \\ \vdots & \vdots & & \\ a_{m1} & a_{m2} & \dots & a_{mn} \end{pmatrix} \begin{pmatrix} u_1 \\ u_2 \\ \vdots \\ u_n \end{pmatrix} = \begin{pmatrix} b_1 \\ b_2 \\ \vdots \\ b_n \end{pmatrix}, \tag{2.93}$$

or simply

$$\mathbf{A}\mathbf{u} = \mathbf{b}. \tag{2.94}$$

In the case of $m = n$, we multiply both sides by $\mathbf{A}^{-1}$ (this is only possible when $m = n$),

$$\mathbf{A}^{-1}\mathbf{A}\mathbf{u} = \mathbf{A}^{-1}\mathbf{b}, \tag{2.95}$$

we obtain the solution

$$\mathbf{u} = \mathbf{A}^{-1}\mathbf{b}. \tag{2.96}$$

A special case of the above equation is when $\mathbf{b} = \lambda\mathbf{u}$, and this becomes an eigenvalue problem. An eigenvalue λ and corresponding eigenvector $\mathbf{v}$ of a square matrix $\mathbf{A}$ satisfy

$$\mathbf{A}\mathbf{v} = \lambda\mathbf{v}, \tag{2.97}$$

or

$$(\mathbf{A} - \lambda\mathbf{I})\mathbf{v} = \mathbf{0}. \tag{2.98}$$

Any nontrivial solution requires

$$\begin{vmatrix} a_{11}-\lambda & a_{12} & \dots & a_{1n} \\ a_{21} & a_{22}-\lambda & \dots & a_{2n} \\ \vdots & \vdots & & \\ a_{n1} & a_{n2} & \dots & a_{nn}-\lambda \end{vmatrix} = 0, \tag{2.99}$$

which is equivalent to

$$\lambda^n + \alpha_{n-1}\lambda^{n-1} + \dots + \alpha_0$$

$$= (\lambda - \lambda_1)(\lambda - \lambda_2)\dots(\lambda - \lambda_n) = 0. \tag{2.100}$$

In general, the characteristic equation has n solutions. Eigenvalues have the interesting connections with the matrix,

$$\mathrm{tr}(\mathbf{A}) = \sum_{i=1}^{n} a_{ii} = \lambda_1 + \lambda_2 + \dots + \lambda_n. \tag{2.101}$$

For a symmetric square matrix, the two eigenvectors for two distinct eigenvalues λ_i and λ_j are orthogonal $\mathbf{v}^{\mathbf{T}}\mathbf{v} = 0$.

Some useful identities involving eigenvalues and inverse of matrices are as follows:

$$(\mathbf{AB...Z})^{-1} = \mathbf{Z}^{-1}...\mathbf{B}^{-1}\mathbf{A}^{-1}, \tag{2.102}$$

$$\mathbf{A}\mathbf{v}_i = \lambda_i \mathbf{v}_i, \qquad \lambda_i = \mathrm{eig}(\mathbf{A}), \tag{2.103}$$

$$\mathrm{eig}(\mathbf{AB}) = \mathrm{eig}(\mathbf{BA}), \tag{2.104}$$

$$\mathrm{tr}(\mathbf{A}) = \sum_i \mathbf{A}_{ii} = \sum_i \lambda_i, \qquad \det(\mathbf{A}) = \Pi_i \lambda_i. \tag{2.105}$$

◇ **Example 2.5:** For a simple 2×2 matrix

$$\mathbf{A} = \begin{pmatrix} 1 & 5 \\ 2 & 4 \end{pmatrix},$$

its eigenvalues can be determined by

$$\begin{vmatrix} 1-\lambda & 5 \\ 2 & 4-\lambda \end{vmatrix} = 0,$$

or

$$(1-\lambda)(4-\lambda) - 2 \times 5 = 0,$$

which is equivalent to

$$(\lambda + 1)(\lambda - 6) = 0.$$

Thus, the eigenvalues are $\lambda_1 = -1$ and $\lambda_2 = 6$. The trace of $\mathbf{A}$ is tr$(\mathbf{A}) = A_{11} + A_{22} = 1 + 4 = 5 = \lambda_1 + \lambda_2$.

In order to obtain the eigenvector for each eigenvalue, we assume

$$\mathbf{v} = \begin{pmatrix} v_1 \\ v_2 \end{pmatrix}.$$

For the eigenvalue $\lambda_1 = -1$, we plug this into

$$|\mathbf{A} - \lambda \mathbf{I}|\mathbf{v} = \mathbf{0},$$

and we have

$$\begin{vmatrix} 1-(-1) & 5 \\ 2 & 4-(-1) \end{vmatrix} \begin{pmatrix} v_1 \\ v_2 \end{pmatrix} = 0,$$

or

$$\begin{vmatrix} 2 & 5 \\ 2 & 5 \end{vmatrix} \begin{pmatrix} v_1 \\ v_2 \end{pmatrix} = 0,$$

which is equivalent to

$$2v_1 + 5v_2 = 0, \qquad \text{or} \qquad v_1 = -\frac{5}{2}v_2.$$

This equation has infinite solutions, each corresponds to the vector parallel to the unit eigenvector. As the eigenvector should be normalized so that its modulus is unity, this additional condition requires

$$v_1^2 + v_2^2 = 1,$$

which means

$$(\frac{-5v_2}{2})^2 + v_2^2 = 1.$$

We have $v_1 = -5/\sqrt{29}$, $v_2 = 2/\sqrt{29}$. Thus, we have the first set of eigenvalue and eigenvector

$$\lambda_1 = -1, \qquad \mathbf{v}_1 = \begin{pmatrix} -\frac{5}{\sqrt{29}} \\ \frac{2}{\sqrt{29}} \end{pmatrix}. \tag{2.106}$$

Similarly, the second eigenvalue $\lambda_2 = 6$ gives

$$\begin{vmatrix} 1-6 & 5 \\ 2 & 4-6 \end{vmatrix} \begin{pmatrix} v_1 \\ v_2 \end{pmatrix} = 0.$$

Using the normalization condition $v_1^2 + v_2^2 = 1$, the above equation has the following solution

$$\lambda_2 = 6, \qquad \mathbf{v}_2 = \begin{pmatrix} \frac{\sqrt{2}}{2} \\ \frac{\sqrt{2}}{2} \end{pmatrix}.$$

◇

2.2.3 Iteration Methods

For a linear system $\mathbf{Au} = \mathbf{b}$, the solution $\mathbf{u} = \mathbf{A}^{-1}\mathbf{b}$ generally involves the inversion of a large matrix. The direct inversion becomes impractical if the matrix is very large (say, if $n > 1000$). Many efficient algorithms have been developed for solving such systems. Gauss elimination and Gauss-Seidel iteration are just two examples. Since this book is an introduction to the finite element analysis, the matrices in our context are usually not very large, and subsequently their inverse can easily done using any decent computer very quickly.

Gauss-Seidel Iteration

Gauss-Seidel iteration method provides an efficient way to solve the linear matrix equation $\mathbf{Au} = \mathbf{b}$ by splitting $\mathbf{A}$ into

$$\mathbf{A} = \mathbf{L} + \mathbf{D} + \mathbf{U}, \tag{2.107}$$

where $\mathbf{L}, \mathbf{D}, \mathbf{U}$ are the lower triangle, diagonal and upper triangle matrices of the $\mathbf{A}$, respectively. The n step iteration is updated by

$$(\mathbf{L} + \mathbf{D})\mathbf{u}^{(n)} = \mathbf{b} - \mathbf{U}\mathbf{u}^{(n-1)}. \tag{2.108}$$

This procedure starts from an initial vector $\mathbf{u}^{(0)}$ (usually, $\mathbf{u}^{(0)} = 0$) stops if a prescribed criterion is reached.

Relaxation Method

The above Gauss-Seidel iteration method is still slow, and the relaxation method provides a more efficient iteration procedure. A popular method is the successive over-relaxation method which consists of two steps

$$\mathbf{v}^{(n-1)} = (\mathbf{L} + \mathbf{D} + \mathbf{U})\mathbf{u}^{(n-1)} - \mathbf{b}, \tag{2.109}$$

and

$$\mathbf{u}^{(n)} = \mathbf{u}^{(n-1)} - \omega(\mathbf{L} + \mathbf{D})^{-1}\mathbf{v}^{(n-1)}, \tag{2.110}$$

where $0 < \omega < 2$ is the overrelaxation parameter. Broadly speaking, a small value of $0 < \omega < 1$ corresponds to under-relaxation with slower convergence while $1 < \omega < 2$ leads to over-relaxation and faster convergence.

2.2.4 Nonlinear Equation

Sometimes, the algebraic equations we meet are nonlinear, and direct inversion is not the best technique. In this case, more elaborate techniques should be used.

Simple Iterations

The nonlinear algebraic equation

$$\mathbf{A}(\mathbf{u})\mathbf{u} = \mathbf{b}(\mathbf{u}), \tag{2.111}$$

or

$$\mathbf{F}(\mathbf{u}) = \mathbf{A}(\mathbf{u})\mathbf{u} - \mathbf{b}(\mathbf{u}) = \mathbf{0}, \tag{2.112}$$

can be solved using a simple iteration technique

$$\mathbf{A}(\mathbf{u}^n)\mathbf{u}^{n+1} = \mathbf{b}(\mathbf{u}^n), \qquad n = 0, 1, 2, ... \tag{2.113}$$

until $\|\mathbf{u}^{n+1} - \mathbf{u}^n\|$ is sufficiently small. Iterations require a starting vector $\mathbf{u}^0$, which is often set to $\mathbf{u}^0 = 0$. This method is also referred to as the successive substitution.

If this simple method does not work, the relaxation method can be used. The relaxation technique first gives a tentative new approximation $\mathbf{u}^*$ from $\mathbf{A}(\mathbf{u}^n)\mathbf{u}^* = \mathbf{b}(\mathbf{u}^n)$, then we use

$$\mathbf{u}^{n+1} = \gamma\mathbf{u}^* + (1-\gamma)\mathbf{u}^n, \qquad \gamma \in (0, 1], \tag{2.114}$$

where γ is a prescribed relaxation parameter.

Newton-Raphson Method

The nonlinear equation (2.112) can also be solved using the Newton-Raphson procedure. We approximate $\mathbf{F}(\mathbf{u})$ by a linear function $\mathbf{R}(\mathbf{u}; \mathbf{u}^n)$ in the vicinity of an existing approximation $\mathbf{u}^n$ to $\mathbf{u}$:

$$\mathbf{R}(\mathbf{u}; \mathbf{u}^n) = \mathbf{F}(\mathbf{u}^n) + \mathbf{J}(\mathbf{u}^n)(\mathbf{u} - \mathbf{u}^n), \;\; \mathbf{J}(\mathbf{u}) = \nabla\mathbf{F}, \tag{2.115}$$

where $\mathbf{J}$ is the Jacobian of $\mathbf{F}(\mathbf{u}) = (F_1, F_2, ..., F_M)^T$. For $\mathbf{u} = (u_1, u_2, ..., u_M)^T$, we have

$$\mathbf{J}_{ij} = \frac{\partial F_i}{\partial u_j}. \tag{2.116}$$

To find the next approximation $\mathbf{u}^{n+1}$ from $\mathbf{R}(\mathbf{u}^{n+1}; \mathbf{u}^n) = 0$, one has to solve a linear system with $\mathbf{J}$ as the coefficient matrix

$$\mathbf{u}^{n+1} = \mathbf{u}^n - \mathbf{J}^{-1}\mathbf{F}(\mathbf{u}^n), \tag{2.117}$$

under a given termination criterion $\|\mathbf{u}^{n+1} - \mathbf{u}^n\| \leq \epsilon$.

Chapter 3

ODEs and Numerical Integration

Most mathematical models in engineering are formulated in terms of differential equations. If the variables or quantities (such as velocity, temperature, pressure) change with other independent variables such as spatial coordinates and time, their relationship can in general be written as a differential equation or even a set of differential equations.

3.1 Ordinary Differential Equations

An ordinary differential equation (ODE) is a relationship between a function $y(x)$ of an independent variable x and its derivatives y', y'', ..., $y^{(n)}$. It can be written in a generic form

$$\Psi(x, y, y', y'', ..., y^{(n)}) = 0. \tag{3.1}$$

The solution of the equation is a function $y = f(x)$, satisfying the equation for all x in a given domain Ω.

The order of the differential equation is equal to the order n of the highest derivative in the equation. Thus, the Riccati equation:

$$y' + a(x)y^2 + b(x)y = c(x), \tag{3.2}$$

is a first order ODE, and the following equation of Euler-type

$$x^2y'' + a_1xy' + a_0y = 0, \tag{3.3}$$

is a second order. The degree of the equation is defined as the power to which the highest derivative occurs. Therefore, both Riccati equation and Euler equation are of the first degree. An equation is called linear if it can be arranged into the form

$$a_n(x)y^{(n)} + ... + a_1(x)y' + a_0(x)y = \phi(x), \tag{3.4}$$

where all the coefficients depend on x only, not on y or any derivatives. If any of the coefficients is a function of y or any of its derivatives, then the equation is nonlinear. If the right hand side is zero or $\phi(x) \equiv 0$, the equation is homogeneous. It is called nonhomogeneous if $\phi(x) \neq 0$.

The solution of an ordinary differential equation is not always straightforward, and it is usually very complicated for nonlinear equations. Even for linear equations, the solutions can only be obtained for a few simple types. The solution of a differential equation generally falls into three types: closed form, series form and integral form. A closed form solution is the type of solution that can be expressed in terms of elementary functions and some arbitrary constants. Series solutions are the ones that can be expressed in terms of a series when a closed-form is not possible for certain type of equations. The integral form of solutions or quadrature is sometimes the only form of solutions that are possible. If all these forms are not possible, the alternatives are to use approximate and numerical solutions.

3.1.1 First Order ODEs

Linear ODEs

The general form of a first order linear differential equation can be written as

$$y' + a(x)y = b(x). \tag{3.5}$$

This equation is always solvable using the integrating factor and it has a closed form solution.

Multiplying both sides of the equation by $\exp[\int a(x)dx]$, which is often called the integrating factor, we have

$$y' e^{\int a(x)dx} + a(x) y e^{\int a(x)dx} = b(x) e^{\int a(x)dx}, \tag{3.6}$$

which can be written as

$$[y e^{\int a(x)dx}]' = b(x) e^{\int a(x)dx}. \tag{3.7}$$

By simple integration, we have

$$y e^{\int a(x)dx} = \int b(x) e^{\int a(x)dx} dx + C. \tag{3.8}$$

So its solution becomes

$$y(x) = e^{-\int a(x)dx} \int b(x) e^{\int a(x)dx} dx + C e^{-\int a(x)dx}, \tag{3.9}$$

where C is an integration constant. The integration constant can be determined if extra requirements are given, and these extra requirements are usually the initial condition when time is zero or boundary conditions at some given points which are at the domain boundary. However, the classification of conditions may also depend on the meaning of the independent x. If x is spatial coordinate, then $y(x = 0) = y_0$ is boundary condition at $x = 0$. However, if $x = t$ means time, then $y(t = 0) = y_0$ can be thought of as the initial condition at $t = 0$. Nevertheless, one integration constant usually requires one condition to determine it.

◇ **Example 3.1:** We now try to solve the ordinary differential equation $\frac{dy}{dt} + ty(t) = -t$ with an initial condition $y(0) = 0$. As $a(t) = t, b(t) = -t$, its general solution is

$$y(t) = e^{-\int t dt} \int (-t) e^{\int t dt} dt + C e^{-\int t dt}$$

$$= -e^{-\frac{t^2}{2}} \int t e^{\frac{t^2}{2}} dt + C e^{-\frac{t^2}{2}}$$

$$= -e^{\frac{t^2}{2}} e^{\frac{t^2}{2}} + Ce^{-\frac{t^2}{2}} = -1 + Ce^{-\frac{t^2}{2}}.$$

From the initial condition $y(0) = 0$ at $t = 0$, we have

$$0 = -1 + C, \qquad \text{or} \qquad C = 1.$$

Thus, the solution becomes

$$y(t) = e^{-\frac{t^2}{2}} - 1.$$

◇

Nonlinear ODEs

For some nonlinear first order ordinary differential equations, sometimes a transform or change of variables can convert it into the standard first order linear equation (3.5). For example, the Bernoulli's equation can be written in the generic form

$$y' + p(x)y = q(x)y^n, \qquad n \neq 1. \tag{3.10}$$

In the case of $n = 1$, it reduces to a standard first order linear ordinary differential equation. By dividing both sides by y^n and using the change of variables

$$u(x) = \frac{1}{y^{n-1}}, \qquad u' = \frac{(1-n)y'}{y^n}, \tag{3.11}$$

we have

$$u' + (1-n)p(x)u = (1-n)q(x), \tag{3.12}$$

which is a standard first order linear differential equation whose general solution is given earlier in this section.

Higher Order ODEs

Higher order ODEs are more complicated to solve even for the linear equations. For the special case of higher-order ODEs where all the coefficients $a_n, ..., a_1, a_0$ are constants,

$$a_n y^{(n)} + ... + a_1 y' + a_0 y = f(x), \tag{3.13}$$

its general solution $y(x)$ consists of two parts: the complementary function $y_c(x)$ and the particular integral or particular solution $y_p^*(x)$. We have

$$y(x) = y_c(x) + y_p^*(x). \tag{3.14}$$

Most nonlinear ordinary differential equations are very difficult to solve, and some higher orders boundary conditions are do not have explicit solutions. In fact, only a few special cases have closed-form solutions. The alternative methods are using the numerical integrations and finite different methods as discussed later in next section.

General Solution

The complementary function is the solution of the linear homogeneous equation with constant coefficients and can be written in a generic form

$$a_n y_c^{(n)} + a_{n-1} y_c^{(n-1)} + ... + a_1 y_c' + a_0 = 0. \tag{3.15}$$

Assuming $y = Ae^{\lambda x}$, we get the polynomial equation of characteristics

$$a_n \lambda^n + a_{n-1} \lambda^{(n-1)} + ... + a_1 \lambda + a_0 = 0, \tag{3.16}$$

which has n roots in general. Then, the solution can be expressed as the summation of various terms $y_c(x) = \sum_{k=1}^{n} c_k e^{\lambda_k x}$ if the polynomial has n distinct zeros $\lambda_1, ...\lambda_n$. For complex roots, and complex roots always occur in pairs $\lambda = r \pm i\omega$, the corresponding linearly independent terms can then be replaced by $e^{rx}[A\cos(\omega x) + B\sin(\omega x)]$.

The particular solution $y_p^*(x)$ is any $y(x)$ that satisfies the original inhomogeneous equation (3.13). Depending on the form of the function $f(x)$, the particular solutions can take various forms. For most of the combinations of basic functions such as $\sin x, \cos x$, e^{kx}, and x^n, the method of the undetermined coefficients is widely used. For $f(x) = \sin(\alpha x)$ or $\cos(\alpha x)$, then we can try $y_p^* = A\sin\alpha x + B\sin\alpha x$. We then substitute it into the original equation (3.13) so that the coefficients A and B can be determined. For a polynomial $f(x) = x^n (n = 0, 1, 2, ..., N)$, we then try $y_p^* = A + Bx + Cx^2 + ... + Qx^n$ (polynomial). For $f(x) = e^{kx}x^n$, $y_p^* = (A + Bx + Cx^2 + ...Qx^n)e^{kx}$. Similarly, $f(x) = e^{kx}\sin\alpha x$ or $f(x) = e^{kx}\cos\alpha x$, we can use $y_p^* = e^{kx}(A\sin\alpha x + B\cos\alpha x)$. More general cases and their particular solutions can be found in various textbooks.

The methods for finding particular integrals work for most cases. However, there are some problems in the case when the right-hand side of the differential equation has the same form as part of the complementary function. In this case, the trial function should include one higher order term obtained by multiplying the standard trial function by the lowest integer power of x so that the product does not appear in any term of the complementary function.

Differential Operator

A very useful technique is to use the method of differential operator D. A differential operator D is defined as

$$D \equiv \frac{d}{dx}. \tag{3.17}$$

Since we know that $De^{\lambda x} = \lambda e^{\lambda x}$ and $D^n e^{\lambda x} = \lambda^n e^{\lambda x}$, so they are equivalent to $D \mapsto \lambda$, and $D^n \mapsto \lambda^n$. Thus, any polynomial $P(D)$ will map to $P(\lambda)$. On the other hand, the integral operator $D^{-1} = \int dx$ is just the inverse of the differentiation. The beauty of using the differential operator form is that one can factorize it in the same way as for factorizing polynomials, then solve each factor separately. Thus, differential operators are very useful in finding out both the complementary functions and particular integral.

◇ **Example 3.2:** To find the particular integral for the equation

$$y''''' + 2y = 17e^{2x},$$

we get

$$(D^5 + 2)y^* = 17e^{2x},$$

or

$$y^* = \frac{17}{D^5+2}e^{2x}.$$

Since $D^5 \mapsto \lambda^5 = 2^5$, we have

$$y^* = \frac{17e^{2x}}{2^5+2} = \frac{e^{2x}}{2}.$$

◇

This method also works for $\sin x, \cos x, \sinh x$ and others, and this is because they are related to $e^{\lambda x}$ via $\sin\theta = \frac{1}{2i}(e^{i\theta} - e^{-i\theta})$ and $\cosh x = (e^x + e^{-x})/2$.

Higher order differential equations can conveniently be written as a system of differential equations. In fact, an nth-order linear equation can always be written as a linear system of n first-order differential equations. A linear system of ODEs is more suitable for mathematical analysis and numerical integration.

3.1.2 Eigenvalue Problems

One of the commonly used second-order ordinary differential equation is the Sturm-Liouville equation in the interval $x \in [a,b]$

$$\frac{d}{dx}[p(x)\frac{dy}{dx}] + q(x)y + \lambda r(x)y = 0, \tag{3.18}$$

with the boundary conditions

$$y(a) + \alpha y'(a) = 0, \qquad y(b) + \beta y'(b) = 0, \tag{3.19}$$

where the known function $p(x)$ is differentiable, and the known functions $q(x), r(x)$ are continuous. The parameter λ to be determined can only take certain values λ_n, called the eigenvalues, if the problem has solutions. For the obvious reason, this problem is called the Sturm-Liouville eigenvalue problem.

For each eigenvalue λ_n, there is a corresponding solution ψ_{λ_n}, called eigenfunctions. The Sturm-Liouville theory states that for two different eigenvalues $\lambda_m \neq \lambda_n$, their eigenfunctions are orthogonal. That is

$$\int_a^b \psi_{\lambda_m}(x)\psi_{\lambda_n}(x)r(x)dx = 0. \tag{3.20}$$

or more generally

$$\int_a^b \psi_{\lambda_m}(x)\psi_{\lambda_n}(x)r(x)dx = \delta_{mn}. \tag{3.21}$$

It is possible to arrange the eigenvalues in an increasing order

$$\lambda_1 < \lambda_2 < .... < \lambda_n < ... \to \infty. \tag{3.22}$$

Sometimes, it is possible to transform a nonlinear equation into a standard linear equation. For example, the Riccati equation can be written in the generic form

$$y' = p(x) + q(x)y + r(x)y^2, \qquad r(x) \neq 0. \tag{3.23}$$

If $r(x) = 0$, then it reduces to a first order linear ODE. By using the transform

$$y(x) = -\frac{u'(x)}{r(x)u(x)}, \tag{3.24}$$

or

$$u(x) = e^{-\int r(x)y(x)dx}, \tag{3.25}$$

we have

$$u'' - P(x)u' + Q(x)u = 0, \tag{3.26}$$

where

$$P(x) = -\frac{r'(x) + r(x)q(x)}{r(x)}, \qquad Q(x) = r(x)p(x). \tag{3.27}$$

As an example, let us look at the buckling of an Euler column which is essentially an elastic rod with one pin-jointed end and the applied axial load P at the other end. The column has a length of L. Its Young's modulus is E and its second moment of area is $I = \int y^2 dA = const$ (for a given geometry). Let $u(x)$ be the transverse displacement, the Euler beam theory gives the following governing equation

$$\frac{EI}{P}\frac{d^2u}{dx^2} + u = 0, \tag{3.28}$$

or

$$u'' + \alpha^2 u = 0, \qquad \alpha^2 = \frac{P}{EI}, \tag{3.29}$$

which is an eigenvalue problem. Its general solution is

$$u = A \sin \alpha x + B \cos \alpha x. \tag{3.30}$$

Applying the boundary conditions, we have at the fixed end

$$u = 0 \qquad (\text{at } \; x = 0), \qquad B = 0, \tag{3.31}$$

and at the free end

$$u = 0, \qquad (\text{at } \; x = L), \qquad A \sin(\alpha L) = 0. \tag{3.32}$$

Thus we have two kinds of solutions either $A = 0$ or $\sin(\alpha L) = 0$. For $A = 0$, we have $u(x) = 0$ which is a trivial solution. So the non-trivial solution requires that

$$\sin(\alpha L) = 0, \tag{3.33}$$

or

$$\alpha L = 0 \text{ (trivial)}, \; \pi, \; 2\pi, \; ..., \; n\pi, \; ... \tag{3.34}$$

Therefore, we have

$$P = \alpha^2 EI = \frac{n^2 \pi^2 EI}{L^2}, \qquad (n = 1, 2, 3, ...). \tag{3.35}$$

The solutions have fixed mode shapes (sine functions) at some critical values (eigenvalues P_n). The lowest eigenvalue is

$$P_* = \frac{\pi^2 EI}{L}, \tag{3.36}$$

which is the Euler buckling load for an elastic rod.

3.2 Finite Difference Scheme

As we have seen earlier in this chapter, it is usually difficult to obtain a closed-form solutions for most differential equations, especially the nonlinear ODEs. In this case, an alternative way is to use numerical methods. The finite difference method is one of the most popular methods that are used commonly in computer simulations. It has the advantage of simplicity and clarity, especially in 1-D configuration and other cases with regular geometry. The finite difference method essentially transforms an ordinary differential equation into a set of algebraic equations by replacing the continuous derivatives with finite difference approximations on a grid of mesh or node points that spans the domain of interest based on the Taylor expansions. In general, the boundary conditions and boundary nodes need special treatment.

3.2.1 Integration of ODEs

The second-order or higher order ordinary differential equations can be written as a first-order system of ODEs. Since the technique for solving a system is essentially the same as that for solving a single equation

$$\frac{dy}{dx} = f(x, y), \tag{3.37}$$

then we shall focus on the first-order equation in the rest of this section. In principle, the solution can be obtained by direct integration,

$$y(x) = y_0 + \int_{x_0}^{x} f(x, y(x))dx, \tag{3.38}$$

but in practice it is usually impossible to do the integration analytically as it requires the solution of $y(x)$ to evaluate the right-hand side. Thus, some approximations shall be utilized. Numerical integration is the most common technique to obtain approximate solutions. There are various integration schemes with different orders of accuracy and convergent rates. These schemes include the simple Euler scheme, Runge-Kutta method, Relaxation method, and many others.

3.2.2 Euler Scheme

Using the notations $h = \Delta x = x_{n+1} - x_n$, $y_n = y(x_n)$, $x_n = x_0 + n\Delta x$ $(n = 0, 1, 2, ..., N)$, and $' = d/dx$ for convenience, then the explicit Euler scheme can simply be written as

$$y_{n+1} = y_n + \int_{x_n}^{x_{n+1}} f(x, y)dx \approx y_n + hf(x_n, y_n). \tag{3.39}$$

This is a forward difference method as it is equivalent to the approximation of the first derivative

$$y'_n = \frac{y_{n+1} - y_n}{\Delta x}. \tag{3.40}$$

The order of accuracy can be estimated using the Taylor expansion

$$y_{n+1} = y_n + hy'|_n + \frac{h^2}{2}y''|_n + ...$$

$$\approx y_n + hf(x_n, y_n) + O(h^2). \tag{3.41}$$

Thus, the Euler method is first order accurate.

For any numerical algorithms, the algorithm must be stable in order to reach convergent solutions. Thus, stability is an important issue in numerical analysis. Defining δy as the discrepancy between the actual numerical solution and the true solution of the Euler finite difference equation, we have

$$\delta y_{n+1} = [1 + hf'(y)] = \xi \delta y_n. \tag{3.42}$$

In order to avoid the discrepancy to grow, it requires the following stability condition $|\xi| \leq 1$. The stability restricts the size of interval h, which is usually small. One alternative that can use larger h is the implicit Euler scheme, and this scheme approximates the derivative by a backward difference $y'_n = (y_n - y_{n-1})/h$ and the right-hand side of equation (3.38) is evaluated at the new y_{n+1} location. Now the scheme can be written as

$$y_{n+1} = y_n + hf(x_{n+1}, y_{n+1}). \tag{3.43}$$

The stability condition becomes

$$\delta y_{n+1} = \xi \delta y_n = \frac{\delta y_n}{1 - hf'(y)}, \tag{3.44}$$

which is always stable if $f'(y) = \frac{\partial f}{\partial y} \leq 0$. This means that any step size is acceptable. However, the step size cannot be too large as the accuracy reduces as the step size increases. Another practical issue is that, for most problems such as non-linear ODEs, the evaluation of y' and $f'(y)$ requires the value of y_{n+1} which is unknown. Thus, an iteration procedure is needed to march to a new value y_{n+1}, and the iteration starts with a guess value which is usually taken to be zero for most cases. The implicit scheme generally gives better stability.

◇ **Example 3.3:** To solve the equation

$$\frac{dy}{dx} = f(y) = e^{-y} - y,$$

we use the explicit Euler scheme, and we have

$$y_{n+1} \approx y_n + hf(y_n) = y_n + h(e^{-y_n} - y_n).$$

Suppose the discrepancy between true solution y_n^* and the numerical y_n is δy_n so that $y_n^* = y_n + \delta y_n$, then the true solution satisfies

$$y_{n+1}^* = y_n^* + hf(y_n^*).$$

Since $f(y_n^*) = f(y_n) + \frac{df}{dy}\delta y_n$, the above equation becomes

$$y_{n+1} + \delta y_{n+1} = y_n + \delta y_n + h[f(y_n) + f'(y_n)\delta y_n].$$

Together with the Euler scheme, we have

$$\delta y_{n+1} = \delta y_n + f'\delta y_n.$$

Suppose that $\delta y_n \propto \xi^n$, then we have

$$\xi^{n+1} = \xi^n + hf'\xi^n, \qquad \text{or} \qquad \xi = 1 + hf'.$$

In order for the scheme to be stable (or $\xi^n \to 0$), it requires that

$$|\xi| \leq 1, \qquad \text{or} \qquad -1 \leq 1 + hf' = 1 - h(e^{-y_n} + 1) \leq 1.$$

The stability condition becomes

$$0 \leq h \leq \frac{2}{e^{-y_n} + 1}.$$

◇

3.2.3 Leap-Frog Method

The Leap-frog scheme is the central difference

$$y_n' = \frac{y_{n+1} - y_{n-1}}{2\Delta x}, \tag{3.45}$$

which leads to

$$y_{n+1} = y_{n-1} + 2hf(x_n, y_n). \tag{3.46}$$

The central difference method is second order accurate. In a similar way as equation (3.42), the leap frog method becomes

$$\delta y_{n+1} = \delta y_{n-1} + 2hf'(y)\delta y_n, \tag{3.47}$$

or

$$\delta y_{n+1} = \xi^2 \delta y_{n-1}, \tag{3.48}$$

where $\xi^2 = 1 + 2hf'(y)\xi$. This scheme is stable only if $|\xi| \leq 1$, and a special case is $|\xi| = 1$ when $f'(y)$ is purely imaginary. Therefore, the central scheme is not necessarily a better scheme than the forward scheme.

3.2.4 Runge-Kutta Method

We have so far seen that stability of the Euler method and the central difference method is limited. The Runge-Kutta method uses a trial step to the midpoint of the interval by central difference and combines with the forward difference at two steps

$$\hat{y}_{n+1/2} = y_n + \frac{h}{2} f(x_n, y_n), \tag{3.49}$$

$$y_{n+1} = y_n + hf(x_{n+1/2}, \hat{y}_{n+1/2}). \tag{3.50}$$

This scheme is second order accurate with higher stability compared with previous simple schemes. One can view this scheme as a predictor-corrector method. In fact, we can use multisteps to devise higher order methods if the right combinations are used to eliminate the error terms order by order. The popular classical Runge-Kutta method can be written as

$$a = hf(x_n, y_n),$$

$$b = hf(x_n + h/2, y_n + a/2),$$

$$c = hf(x_n + h, y_n + b/2),$$

$$d = hf(x_n + h, y_n + c),$$

$$y_{n+1} = y_n + \frac{a + 2(b + c) + d}{6}, \tag{3.51}$$

which is fourth order accurate. Generally speaking, the higher-order scheme is better than the lower scheme, but not always.

Chapter 4

PDEs and Finite Difference

Partial differential equations (PDEs) are much more complicated compared with the ordinary differential equations. There is no universal solution technique for nonlinear equations, even the numerical simulations are usually not straightforward. Thus, we will mainly focus on the linear partial differential equations and the equations of special interests in engineering and computational sciences. A partial differential equation (PDE) is a relationship containing one or more partial derivatives. Similar to the ordinary differential equation, the highest nth partial derivative is referred to as the order n of the partial differential equation. The general form of a partial differential equation can be written as

$$\psi(x, y, ..., \frac{\partial u}{\partial x}, \frac{\partial u}{\partial y}, \frac{\partial^2 u}{\partial x^2}, \frac{\partial^2 u}{\partial y^2}, \frac{\partial^2 u}{\partial x \partial y}, ...) = 0. \tag{4.1}$$

where u is the dependent variable and $x, y, ...$ are the independent variables.

A simple example of partial differential equations is the linear first order partial differential equation, which can be written as

$$a(x, y)\frac{\partial u}{\partial x} + b(x, y)\frac{\partial u}{\partial y} = f(x, y). \tag{4.2}$$

for two independent variables and one dependent variable u. If the right hand side is zero or simply $f(x, y) = 0$, then the equation is said to be homogeneous. The equation is said to be linear if a, b and f are functions of x, y only, not u itself.

For simplicity in notations in the studies of partial differential equations, compact subscript forms are often used in the literature. They are

$$u_x \equiv \frac{\partial u}{\partial x}, \quad u_y \equiv \frac{\partial u}{\partial y}, \quad u_{xx} \equiv \frac{\partial^2 u}{\partial x^2},$$

$$u_{yy} \equiv \frac{\partial^2 u}{\partial y^2}, \quad u_{xy} \equiv \frac{\partial^2 u}{\partial x \partial y}, \quad \dots \tag{4.3}$$

and thus we can write (4.2) as

$$a u_x + b u_y = f. \tag{4.4}$$

In the rest of the chapters in this book, we will use these notations whenever no confusion occurs.

4.1 PDEs and Classification

4.1.1 First-Order PDEs

The first order partial differential equation of linear type can be written as

$$a(x, y) u_x + b(x, y) u_y = f(x, y), \tag{4.5}$$

which can be solved using the method of characteristics

$$\frac{dx}{a} = \frac{dy}{b} = \frac{du}{f}. \tag{4.6}$$

This is equivalent to the following equation in terms of parameter s

$$\frac{dx}{ds} = a, \quad \frac{dy}{ds} = b, \quad \frac{du}{ds} = f, \tag{4.7}$$

which essentially forms a system of first-order ordinary differential equations.

The simplest example of a first order linear partial differential equation is the first order hyperbolic equation

$$u_t + cu_x = 0, \tag{4.8}$$

where c is a constant. It has a general solution of

$$u = \psi(x - ct), \tag{4.9}$$

which is a travelling wave along x-axis with a constant speed c. If the initial shape is $u(x, 0) = \psi(x)$, then $u(x, t) = \psi(x - ct)$ at time t, therefore the shape of the wave does not change though its position is constantly changing.

4.1.2 Classification

A linear second-order partial differential equation can be written in the generic form in terms of two independent variables x and y,

$$au_{xx} + bu_{xy} + cu_{yy} + gu_x + hu_y + ku = f, \tag{4.10}$$

where a, b, c, g, h, k and f are functions of x and y only. If $f(x, y, u)$ is also a function of u, then we say that this equation is quasi-linear.

If $\Delta = b^2 - 4ac < 0$, the equation is elliptic. One famous example is the Laplace equation $u_{xx} + u_{yy} = 0$.

If $\Delta > 0$, it is hyperbolic. One example is the wave equation $u_{tt} = c^2 u_{xx}$.

If $\Delta = 0$, it is parabolic. Diffusion and heat conduction equations are of the parabolic type $u_t = \kappa u_{xx}$.

4.2 Classic PDEs

Many physical processes in engineering are governed by three classic partial differential equations so they are widely used in a vast range of applications.

4.2.1 Laplace's and Poisson's Equation

In heat transfer problems, the steady state of heat conduction with a source is governed by the Poison equation

$$k\nabla^2 u = f(x, y, t), \qquad (x, y) \in \Omega, \tag{4.11}$$

or

$$u_{xx} + u_{yy} = q(x, y, t), \tag{4.12}$$

for two independent variables x and y. Here k is the thermal diffusivity and $f(x, y, t)$ is the heat source. If there is no heat source ($q = 0$), this becomes the Laplace equation. The solution or a function is said to be harmonic if it satisfies the Laplace equation.

4.2.2 Diffusion Equation

Time-dependent problems, such as diffusion and transient heat conduction, are governed by parabolic equations. The heat conduction equation or diffusion equation

$$u_t = k u_{xx}, \tag{4.13}$$

is a famous example. For diffusion problem, k is replaced by the diffusion coefficient D.

4.2.3 Wave Equation

The vibrations of strings and travelling sound waves are governed by the hyperbolic wave equation. The 1-D wave equation in its simplest form is

$$u_{tt} = c^2 u_{xx}, \tag{4.14}$$

where c is the speed of the wave.

There are other equations such as the reaction-diffusion equation that occur frequently in mathematical physics, engineering and computational sciences. We will give a brief description when we meet these equations in later chapters.

4.3 Finite Difference Methods

4.3.1 Hyperbolic Equations

Numerical solutions of partial differential equations are more complicated than that of ODEs because it involves time and space variables and the geometry of the domain of interest. Usually, boundary conditions are more complex. In addition, nonlinear problems are very common in engineering applications. Now we start with the simplest first order equations and then move onto more complicated cases.

First-Order Hyperbolic Equation

For simplicity, we start with the one-dimensional scalar equation of hyperbolic type,

$$\frac{\partial u}{\partial t} + c\frac{\partial u}{\partial x} = 0, \tag{4.15}$$

where c is a constant or the velocity of advection. By using the forward Euler scheme for time and centered-spaced scheme, we have

$$\frac{u_j^{n+1} - u_j^n}{\Delta t} + c[\frac{u_{j+1}^n - u_{j-1}^n}{2h}] = 0, \tag{4.16}$$

where $t = n\Delta t, n = 0, 1, 2, ..., x = x_0 + jh, j = 0, 1, 2, ...,$ and $h = \Delta x$. In order to see how this method behaves numerically, we use the von Neumann stability analysis.

Assuming the independent solutions or eigenmodes (also called Fourier modes) in spatial coordinate x in the form of $u_j^n = \xi^n e^{ikhj}$, and substituting into equation (4.16), we have

$$\xi^{n+1}e^{ikhj} - \xi^n e^{ikhj} = \xi^n \frac{c\Delta t}{h} \frac{e^{ikh(j+1)} - e^{ihk(j-1)}}{2}. \tag{4.17}$$

Dividing both sides of the above equation by $\xi^n \exp(ikhj)$ and using $\sin x = (e^{ix} - e^{-ix})/2i$, we get

$$\xi = 1 - i\frac{c\Delta t}{h}\sin(kh). \tag{4.18}$$

The stability criteria $|\xi| \leq 1$ require

$$(\frac{c\Delta t}{h})^2 \sin^2 kh \leq 0. \tag{4.19}$$

However, this inequality is impossible to satisfy and this scheme is thus unconditionally unstable.

Second-Order Wave Equation

Higher order equations such as the second-order wave equation can be written as a system of hyperbolic equations and then be solved using numerical integration. They can also be solved by direct discretization using a finite difference scheme. The wave equation

$$\frac{\partial^2 u}{\partial t^2} = c^2 \frac{\partial^2 u}{\partial x^2}, \tag{4.20}$$

consists of second derivatives. If we approximate the first derivatives at each time step n using

$$u_i' = \frac{u_{i+1}^n - u_i^n}{\Delta x}, \qquad u_{i-1}' = \frac{u_i^n - u_{i-1}^n}{\Delta x}, \tag{4.21}$$

then we can use the following approximation for the second derivative

$$u_i'' = \frac{u_i' - u_{i-1}'}{\Delta x}$$

$$= \frac{u_{i+1}^n - 2u_i^n + u_{i-1}^n}{(\Delta x)^2}. \tag{4.22}$$

This is in fact a central difference scheme of second order accuracy. If we use the similar scheme for time-stepping, then we get a central difference scheme in both time and space.

Thus, the numerical scheme for this equation becomes

$$\frac{u_i^{n+1} - 2u_i^n + u_i^{n-1}}{(\Delta t)^2} = c^2 \frac{u_{i+1}^n - 2u_i^n + u_{i-1}^n}{(\Delta x)^2}. \tag{4.23}$$

This is a two-level scheme with a second order accuracy. The idea of solving this difference equation is to express (or to solve) u_i^{n+1} at time step $t = n+1$ in terms of the known values or data u_i^n and u_i^{n-1} at two previous time steps $t = n$ and $t = n - 1$.

4.3.2 Parabolic Equation

For the parabolic equation such as the diffusion or heat conduction equation

$$\frac{\partial u}{\partial t} = \frac{\partial}{\partial x}(D\frac{\partial u}{\partial x}), \tag{4.24}$$

a simple Euler method for the time derivative and centered second-order approximations for space derivatives lead to

$$u_j^{n+1} = u_j^n + \frac{D\Delta t}{h^2}(u_{j+1}^n - 2u_j^n + u_{j-1}^n). \tag{4.25}$$

The stability requirement $\xi \leq 1$ leads to the constraint on the time step (see the example),

$$\Delta t \leq \frac{h^2}{2D}. \tag{4.26}$$

This scheme is shown in Figure 4.1 and it is conditionally stable.

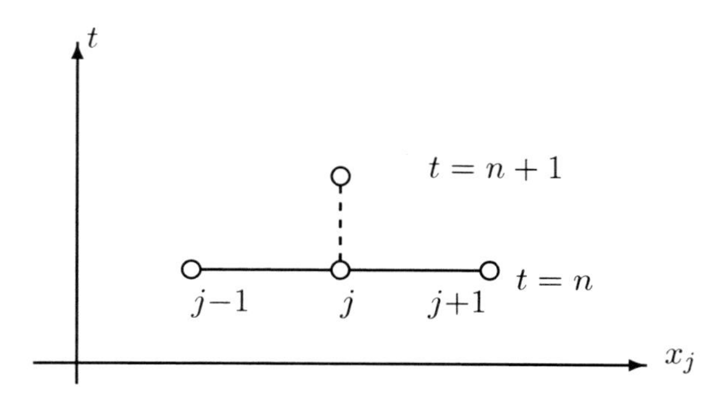

Figure 4.1: Central difference in space and explicit Euler time-stepping.

◇ **Example 4.1:** From equation (4.25), we can apply the von Neumann stability analysis by assuming $u_j^n = \xi^n e^{ikhj}$, we have

$$\xi^{n+1} e^{ikhj} = \xi^n e^{ikhj} + \frac{D\Delta x}{h^2}\xi^n[e^{ikh(j+1)} - 2e^{ikhj} + e^{ikh(j-1)}].$$

Dividing both sides by $\xi^n e^{ikhj}$, we have

$$\xi = 1 + \frac{D\Delta t}{h^2}[e^{ikh} + e^{-ikh} - 2].$$

Using $\cos x = (e^{ix} + e^{-ix})/2$ and $\sin^2(x/2) = (1 - \cos x)/2$, we obtain

$$\xi = 1 - \frac{4D\Delta t}{h^2} \sin^2(\frac{kh}{2}).$$

Since $\sin(x) \le 1$, thus $\xi \le 1$ requires

$$-1 \le 1 - \frac{4D\Delta t}{h^2} \le 1,$$

or

$$0 \le \Delta t \le \frac{h^2}{2D}.$$

◇

A typical feature of a solution to the diffusive system is that the profile is gradually smoothed out as time increases. The time-stepping scheme we used limits the step size of time as larger time steps will make the scheme unstable. There are many ways to improve this, and one of most widely used schemes is the implicit scheme.

To avoid the difficulty caused by very small timesteps, we now use an implicit scheme for time derivative differencing, and thus we have

$$u_j^{n+1} - u_j^n = \frac{D\Delta t}{h^2}(u_{j+1}^{n+1} + 2u_j^{n+1} + u_{j-1}^{n+1}). \tag{4.27}$$

Applying the stability analysis, we have

$$\xi = \frac{1}{1 + \frac{4D\Delta t}{h^2} \sin^2 \frac{kh}{2}}, \tag{4.28}$$

whose norm is always less than unity ($|\xi| \le 1$). This means the implicit scheme is unconditionally stable for any size of time steps. That is why implicit methods are more desired in simulations. However, there is one disadvantage of this method, which requires more programming skills because the inverse of a large matrix is usually needed in implicit schemes.

4.3.3 Elliptical Equation

In the parabolic equation, if the time derivative is zero or u does not change with time $u_t = 0$, then it becomes a steady-state problem that is governed by the elliptic equation. For the steady state heat conduction problem, we generally have the Poisson problem,

$$\nabla \cdot [\kappa(u, x, y, t)\nabla u] = f, \tag{4.29}$$

If κ is a constant, this becomes

$$\nabla^2 u = q, \qquad q = \frac{f}{\kappa}. \tag{4.30}$$

There are many methods available to solve this problems such as the boundary integral method, the relaxation method, and the multigrid method. Two major ones are the long-time approximation of the transient parabolic diffusion equations, the other includes the iteration method.

The long time approximation method is essentially based on fact that the parabolic equation

$$\frac{\partial u}{\partial t} + \kappa \nabla^2 u = f, \tag{4.31}$$

evolves with a typical scale of $\sqrt{\kappa t}$. If $\sqrt{\kappa t} \gg 1$, the system is approaching its steady state. Assuming $t \to \infty$ and $\kappa \gg 1$, we then have

$$\nabla^2 u = \frac{f}{\kappa} - \frac{1}{\kappa} u_t \to 0. \tag{4.32}$$

The long-time approximation is based on the fact that the parabolic equation in the case of $\kappa = \text{const}$ degenerates into the above steady-state equation (4.29) because $u_t \to 0$ as $t \to \infty$. This approximation becomes better if $\kappa \gg 1$. Thus, the usual numerical methods for solving parabolic equations are valid. However, other methods may obtain the results more quickly.

The iteration method uses the second-order scheme for space derivatives, and equation (4.30) in the 2-D case becomes

$$\frac{u_{i+1,j} - 2u_{i,j} + u_{i-1,j}}{(\Delta x)^2} + \frac{u_{i,j+1} - 2u_{i,j} + u_{i,j-1}}{(\Delta y)^2} = q. \tag{4.33}$$

If we use $\Delta x = \Delta y = h$, then the above equation simply becomes

$$(u_{i,j+1} + u_{i,j-1} + u_{i+1,j} + u_{i-1,j}) - 4u_{i,j} = h^2 q, \tag{4.34}$$

which can be written as

$$\mathbf{Au} = \mathbf{b}. \tag{4.35}$$

In principle, one can solve this equation using Gauss elimination; however, this becomes impractical as the matrix becomes large such as 1000×1000. The Gauss-Seidel iteration provides a more efficient way to solve this equation by splitting $\mathbf{A}$ as

$$\mathbf{A} = \mathbf{L} + \mathbf{D} + \mathbf{U}, \tag{4.36}$$

where $\mathbf{L}, \mathbf{D}, \mathbf{U}$ are the lower triangle, diagonal and upper triangle matrices of $\mathbf{A}$, respectively. The iteration is updated in the following way:

$$\mathbf{u^{(n)}} = (\mathbf{D} + \mathbf{L})^{-\mathbf{1}}[\mathbf{b} - \mathbf{U}\mathbf{u^{(n-1)}}]. \tag{4.37}$$

This procedure stops until a prescribed error or precision is reached.

4.4 Pattern Formation

4.4.1 Travelling Wave

The nonlinear reaction-diffusion equation

$$\frac{\partial u}{\partial t} = D\frac{\partial^2 u}{\partial x^2} + f(u), \tag{4.38}$$

can have the travelling wave solution under appropriate conditions of $f(0) = f(1) = 0$, $f(u) > 0$ for $u \in [0, 1]$, and $f'(0) > 0$. For example, $f(u) = \gamma u(1-u)$ satisfies these conditions, and the equation in this special case is called the Kolmogorov-Petrovskii-Piskunov (KPP) equation. By assuming that the travelling wave solution has the form $u(\zeta)$ and $\zeta = x - vt$, and substituting into the above equation, we have

$$Du''(\zeta) + vu'(\zeta) + f(u(\zeta)) = 0. \tag{4.39}$$

This is a second-order ordinary differential equation that can be solved with the appropriate boundary conditions

$$u(-\infty) \to 1, \qquad u(\infty) \to 0. \tag{4.40}$$

The KPP theory suggests that the limit of the speed of the travelling wave satisfies

$$v \geq 2\sqrt{Df'(0)}. \tag{4.41}$$

4.4.2 Pattern Formation

One of the most studied nonlinear reaction-diffusion equations in the 2-D case is the Kolmogorov-Petrovskii-Piskunov (KPP) equation

$$\frac{\partial u}{\partial t} = D(\frac{\partial^2 u}{\partial x^2} + \frac{\partial^2 u}{\partial y^2}) + \gamma q(u), \tag{4.42}$$

and

$$q(u) = u(1-u). \tag{4.43}$$

The KPP equation can describe a huge number of physical, chemical and biological phenomena. The most interesting feature of this nonlinear equation is its ability of generating beautiful patterns. We can solve it using the finite difference scheme by applying the periodic boundary conditions and using a random initial condition $u = \text{random}(n, n)$ where n is the size of the grid.

Figure 4.2 shows the pattern formation of the above equation on a 200×200 grid for $D = 0.2$ and $\gamma = 0.5$. We can see that rings and thin curves are formed, arising from the random initial condition. The landscape surface shows the variations in the location and values of the field $u(x, y)$ can be easily demonstrated.

The following simple 15-line Matlab program can be used to solve this nonlinear system.

```
% ---------------------------------------------
% Pattern formation:  a 15 line matlab program
```

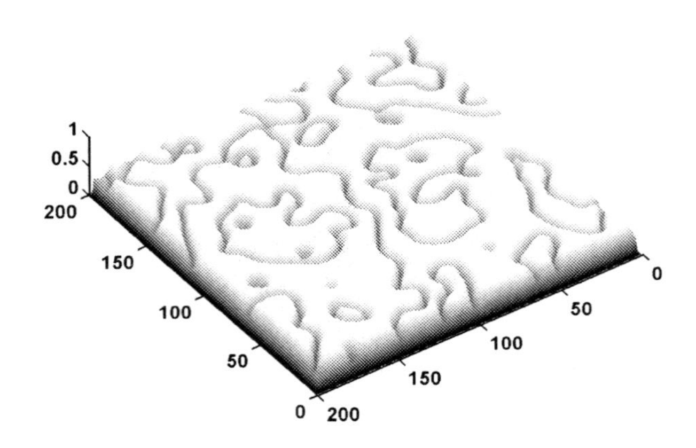

Figure 4.2: 2-D pattern formation for $D = 0.2$ and $\gamma = 0.5$.

```
% PDE form: u_t=D*(u_{xx}+u_{yy})+gamma*q(u)
% where q(u)='u.*(1-u)';
% The solution of this PDE is obtained by the
% finite difference method, assuming dx=dy=dt=1.
% ---------------------------------------------------
% Written by X S Yang (Cambridge University)
% Usage: pattern(100)   or  simply >pattern
% ---------------------------------------------------

function pattern(n)                         % line 1
% Input number of time steps
if nargin<1, n=200; end                     % line 2

% ---------------------------------------------------
% Initialize parameters
% ---- time=100, D=0.2; gamma=0.5; --------------
time=100; D=0.2; gamma=0.5;                 % line 3

% ---- Set initial values of u randomly ---------
u=rand(n,n);  grad=u*0;                     % line 4
```

```
% Vectorisation/index for u(i,j) and the loop ---
I = 2:n-1; J = 2:n-1;                        % line 5

% ------------------------------------------------
% ---- Time stepping ----------------------------
for step=1:time,                              % line 6
% Laplace gradient of the equation           % line 7
 grad(I,J)= u(I,J-1)+u(I,J+1)+u(I-1,J)+u(I+1,J);
 u =(1-4*D)*u+D*grad+gamma*u.*(1-u);         % line 8
% ----- Show results ----------------------------
 pcolor(u);  shading interp;                  % line 9
% ----- Coloring and showing colorbar -----------
 colorbar; colormap jet;                      % line 10
 drawnow;                                     % line 11
end                                           % line 12

% ----- Topology of the final surface ----------
surf(u);                                      % line 13
shading interp;                               % line 14
view([-25 70]);                               % line 15
% ------------- End of Program ----------------
```

If you use this program to do the simulations, you will see that the pattern emerges naturally from the initially random background. Once the pattern is formed, it evolves gradually with time, but the characteristics such as the shape and structure of the patterns do not change much with time. In this sense, one can see beautiful and stable patterns.

4.4.3 Reaction-Diffusion System

The pattern formation in the previous section arises naturally from a single equation of nonlinear reaction-diffusion type. In many applications, we often have to simulate a system of nonlinear reaction-diffusion equations, and the variables are coupled in a complicated manner.

The pattern formation in the previous section comes from

the instability of the nonlinear reaction diffusion system. In order to show this, let us use the following mathematical model for enzyme inhibition and cooperativity.

◇ **Example 4.2:** The following system consists of two nonlinear equations

$$\frac{\partial u}{\partial t} = D_u(\frac{\partial^2 u}{\partial x^2} + \frac{\partial^2 u}{\partial y^2}) + f(u,v),$$

$$\frac{\partial v}{\partial t} = D_v(\frac{\partial^2 v}{\partial x^2} + \frac{\partial^2 v}{\partial y^2}) + g(u,v),$$

and

$$f(u,v) = au(1-u) - \frac{bu - cv}{1+u+v},$$

$$g(u,v) = -\frac{vd}{1+u+v},$$

where D_u and D_v are diffusion coefficients, while a, b, c, d are all constants. This reaction diffusion system may have instability if

$$\frac{D_v}{D_u} < \frac{d}{(2au_0 + b)},$$

where $u_0 = [\sqrt{(b^2 + 4a^2)} - b]/(2a)$. For detailed derivations, see the following stability analysis. ◇

The steady state solutions are obtained from $f(u_0, v_0) = 0$ and $g(u_0, v_0) = 0$. They are

$$u_0 = \frac{b}{2a}[\sqrt{1 + 4\frac{a^2}{b^2}} - 1], \qquad v_0 = 0. \tag{4.44}$$

Let $\psi = (u - u_0, v - v_0)$ be the small perturbation, then ψ satisfies

$$\frac{\partial \psi}{\partial t} = D\nabla^2 \psi + M\psi, \tag{4.45}$$

where

$$D = \begin{pmatrix} D_u & 0 \\ 0 & D_v \end{pmatrix}, \tag{4.46}$$

and

$$M = \frac{1}{(1+u_0)} \begin{pmatrix} -(2au_0 + b) & a(1-u_0) + c \\ 0 & d \end{pmatrix}. \tag{4.47}$$

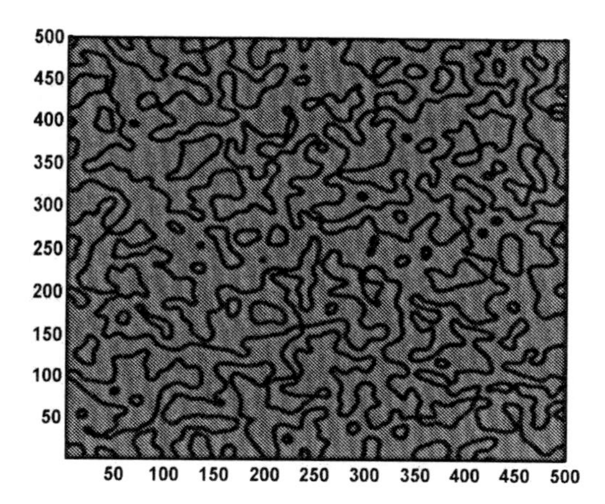

Figure 4.3: 2-D pattern formation for u with $D_a = 0.2, D_b = 0.1$, $\gamma = 0.5$, and $\beta = 0.2$.

Writing ψ in the form of

$$\psi = \sum e^{\lambda t} \psi_k, \tag{4.48}$$

where the summation is over all the wavenumbers k, we have

$$|M - \lambda I - Dk^2| = 0, \tag{4.49}$$

where I is a 2×2 unity matrix. This eigenvalue equation has two roots. Since $\Re(\lambda) > 0$ implies that instability, this requires that

$$\frac{D_v}{D_u} < \frac{d}{(2au_0 + b)}. \tag{4.50}$$

The range of unstable wavenumbers between the two roots of k^2 at the bifurcation point is given by

$$k_{\pm}^2 = \frac{dD_u - D_v(2au_0 + b)}{2D_u D_v(1 + u_0)} [1 \pm \sqrt{1 + 4D_u D_v \delta}], \tag{4.51}$$

with

$$\delta = \frac{(2au_0 + b)}{[dD_u - D_v(2au_0 + b)]^2}. \tag{4.52}$$

If the unstable criteria are satisfied, any small random perturbation can generate complex patterns.

Similar to the nonlinear KPP equation (4.43), beautiful patterns (see Fig. 4.3) also arise naturally in the following nonlinear system

$$\frac{\partial u}{\partial t} = D_a(\frac{\partial^2 u}{\partial x^2} + \frac{\partial^2 u}{\partial y^2}) + \gamma \tilde{f}(u, v), \tag{4.53}$$

$$\frac{\partial v}{\partial t} = D_b(\frac{\partial^2 v}{\partial x^2} + \frac{\partial^2 v}{\partial y^2}) + \beta \tilde{g}(u, v), \tag{4.54}$$

and

$$\tilde{f}(u, v) = u(1 - u), \qquad \tilde{g}(u, v) = u - u^2 v, \tag{4.55}$$

for the values $D_a = 0.2$, $D_b = 0.1$, $\gamma = 0.5$ and $\beta = 0.2$. With different functions $\tilde{f}(u, v)$ and $\tilde{g}(u, v)$, these equations can be used to simulate the pattern formation in a wide range of applications where nonlinear reaction-diffusion equations are concerned.

Chapter 5

Calculus of Variations

The calculus of variations is important in many optimization problems and computational sciences, especially in the formulation of the finite element methods. In this chapter, we will briefly touch these topics.

The main aim of the calculus of variations is to find a function that makes the integral stationary, making the value of the integral a local maximum or minimum. For example, in mechanics we may want to find the shape $y(x)$ of a rope or chain when suspended under its own weight from two fixed points. In this case, the calculus of variations provides a method for finding the function $y(x)$ so that the curve $y(x)$ minimizes the gravitational potential energy of the hanging rope system.

5.1 Euler-Lagrange Equation

5.1.1 Curvature

Before we proceed to the calculus of variations, let us first discuss an important concept, namely the curvature of a curve. In general, a curve $y(x)$ can be described in a parametric form in terms of a vector $\mathbf{r}(s)$ with a parameter s which is the arc length along the curve measured from a fixed point. The curvature κ of a curve is defined as the rate at which the unit tangent $\mathbf{t}$ changes with respect to s. The change of arc length is

$$\frac{ds}{dx} = \sqrt{1 + (\frac{dy}{dx})^2} = \sqrt{1 + y'^2}. \tag{5.1}$$

We have the curvature

$$\frac{d\mathbf{t}}{ds} = \kappa\, \mathbf{n} = \frac{1}{\rho}\mathbf{n}, \tag{5.2}$$

where ρ is the radius of the curvature, and $\mathbf{n}$ is the principal normal. As the direction of the tangent is defined by the angle θ made with the x-axis by $\mathbf{t}$, we have $\tan\theta = y'$. Hence, the curvature becomes

$$\kappa = \frac{d\theta}{ds} = \frac{d\theta}{dx}\frac{dx}{ds}. \tag{5.3}$$

From $\theta = \tan^{-1} y'(x)$, we have

$$\frac{d\theta}{dx} = [\tan^{-1}(y')]' = \frac{y''}{(1 + y'^2)}. \tag{5.4}$$

Using the expression for ds/dx, the curvature can be written in terms of $y(x)$, and we get

$$\kappa = |\frac{d^2\mathbf{r}}{ds^2}| = \frac{y''}{[1 + (y')^2]^{3/2}}. \tag{5.5}$$

5.1.2 Euler-Lagrange Equation

Since the calculus of variations is always related to some minimization or maximization, we can in general assume that the integrand ψ of the integral is a function of the shape or curve $y(x)$ (shown in Figure 5.1), its derivative $y'(x)$ and the spatial coordinate x (or time t, depending on the context). For the integral

$$I = \int_a^b \psi(x, y, y')dx, \tag{5.6}$$

where a and b are fixed, the aim is to find the solution of the curve $y(x)$ such that it makes the value of I stationary. In this

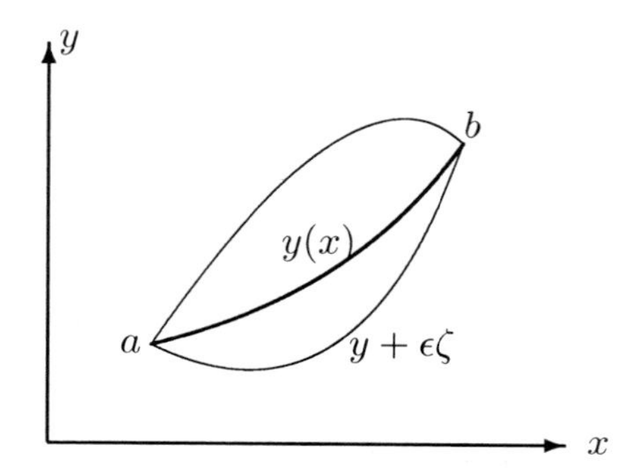

Figure 5.1: Variations in the path $y(x)$.

sense, $I[y(x)]$ is a function of the function $y(x)$, and thus it is referred to as the functional.

Here, stationary means that the small change of the first order in $y(x)$ will only lead to the second-order changes in values of $I[y(x)]$, and subsequently, the change δI of I should be virtually zero due to the small variation in the function $y(x)$. Translating this into the mathematical language, we suppose that $y(x)$ has a small change of magnitude of ϵ so that

$$y(x) \rightarrow y(x) + \epsilon\zeta(x), \tag{5.7}$$

where $\zeta(x)$ is an arbitrary function. The requirement of I to be stationary means that

$$\delta I = 0, \tag{5.8}$$

or more accurately,

$$\frac{dI}{d\epsilon}|_{\epsilon=0} = 0, \qquad \text{for} \quad \text{all} \quad \zeta(x). \tag{5.9}$$

Thus I becomes

$$I(y, \epsilon) = \int_a^b \psi(x, y + \epsilon\zeta, y' + \epsilon\zeta')dx$$

$$= \int_a^b \psi(x,y,y')dx + \int_a^b [\epsilon(\zeta\frac{\partial \psi}{\partial y} + \zeta'\frac{\partial \psi}{\partial y'}]dx + O(\epsilon^2). \quad (5.10)$$

The first derivative of I should be zero, and we have

$$\frac{\delta I}{\delta \epsilon} = \int_a^b [\frac{\partial \psi}{\partial y}\zeta + \frac{\partial \psi}{\partial y'}\zeta']dx = 0, \quad (5.11)$$

which is exactly what we mean that the change δI (or the first order variation) in the value of I should be zero. Integrating this equation by parts, we have

$$\int_a^b [\frac{\partial \psi}{\partial y} - \frac{d}{dx}\frac{\partial \psi}{\partial y'}]\zeta dx = -[\zeta\frac{\partial \psi}{\partial y'}]\Big|_a^b. \quad (5.12)$$

If we require that $y(a)$ and $y(b)$ are known at the fixed points $x = a$ and $x = b$, then these requirements naturally lead to $\zeta(a) = \zeta(b) = 0$. This means that the above right hand side of the equation is zero. That is,

$$\Big[\zeta\frac{\partial \psi}{\partial y'}\Big]_a^b = 0, \quad (5.13)$$

which gives

$$\int_a^b \Big[\frac{\partial \psi}{\partial y} - \frac{d}{dx}\frac{\partial \psi}{\partial y'}\Big]\zeta dx = 0. \quad (5.14)$$

As this equation holds for all $\zeta(x)$, the integrand must be zero. Therefore, we have the well-known Euler-Lagrange equation

$$\frac{\partial \psi}{\partial y} = \frac{d}{dx}(\frac{\partial \psi}{\partial y'}). \quad (5.15)$$

It is worth pointing out that this equation is very special in the sense that ψ is known and the unknown is $y(x)$. It has many applications in mathematics, natural sciences and engineering.

The simplest and classical example is to find the shortest path on a plane joining two points, say, $(0,0)$ and $(1,1)$. We know that the total length along a curve $y(x)$ is

$$L = \int_0^1 \sqrt{1 + y'^2}dx. \quad (5.16)$$

Since $\psi = \sqrt{1+y'^2}$ does not contain y, thus $\frac{\partial \psi}{\partial y} = 0$. From the Euler-Lagrange equation, we have

$$\frac{d}{dx}\left(\frac{\partial \psi}{\partial y'}\right) = 0, \tag{5.17}$$

its integration is

$$\frac{\partial \psi}{\partial y'} = \frac{y'}{\sqrt{1+y'^2}} = A. \tag{5.18}$$

Rearranging it as

$$y'^2 = \frac{A^2}{1-A^2}, \qquad \text{or} \qquad y' = \frac{A}{\sqrt{1-A^2}}, \tag{5.19}$$

and integrating again, we have

$$y = kx + c, \qquad k = \frac{A}{\sqrt{1-A^2}}. \tag{5.20}$$

This is a straight line. That is exactly what we expect from the plane geometry.

◇ **Example 5.1:** The Euler-Lagrange equation is very general and includes many physical laws if the appropriate form of ψ is used. For a point mass m following under the Earth's gravity g, the action (see below) is defined as

$$\psi = \frac{1}{2}mv^2 - mgy = \frac{1}{2}m(\dot{y})^2 - mgy,$$

where $y(t)$ is the path, and now x is replaced by t. $v = \dot{y}$ is the velocity. The Euler-Lagrange equation becomes

$$\frac{\partial \psi}{\partial y} = \frac{d}{dt}\left(\frac{\partial \psi}{\partial v}\right),$$

or

$$-mg = \frac{d}{dt}(mv),$$

which is essentially the Newton's second law $F = ma$ because the right hand side is the rate of change of the momentum mv, and the left hand side is the force. ◇

Well, you may say, this is trivial and there is nothing new about it. This example is indeed too simple. Let us now study a more complicated case so as to demonstrate the wide applications of the Euler-Lagrange equation. In mechanics, there is a Hamilton's principle which states that the configuration of a mechanical system is such that the action integral I of the Lagrangian $\mathcal{L} = T - V$ is stationary with respect to the variations in the path. That is to say that the configuration can be uniquely defined by its coordinates q_i and time t, when moving from one configuration at time t_0 to another time $t = t*$

$$I = \int_0^{t^*} \mathcal{L}(t, q_i, \dot{q}_i) dt, \qquad i = 1, 2, ..., N, \tag{5.21}$$

where T is the total kinetic energy (usually, a function of $\dot{q}_i$), and V is the potential energy (usually, a function of q). Here $\dot{q}_i$ means

$$\dot{q}_i = \frac{\partial q_i}{\partial t}. \tag{5.22}$$

In analytical mechanics and engineering, the Lagrangian $\mathcal{L}$ (=kinetic energy - potential energy) is often called the action, thus this principle is also called the principle of least action. The physical configuration or the path of movement follows such a path that makes the action integral stationary.

In the special case, $x \to t$, the Euler-Lagrange equation becomes

$$\frac{\partial \mathcal{L}}{\partial q_i} = \frac{d}{dt}\left(\frac{\partial \mathcal{L}}{\partial \dot{q}_i}\right), \tag{5.23}$$

which is the well-known Lagrange's equation. This seems too abstract. Now let us look at a classic example.

◇ **Example 5.2:** For a simple pendulum shown in Figure 5.2, we now try to derive its equation of oscillations. We know the kinetic energy T and the potential energy V are

$$T = \frac{1}{2} m l^2 \left(\frac{d\theta}{dt}\right)^2 = \frac{1}{2} m l^2 \dot{\theta}^2, \qquad V = mgh = mgl(1 - \cos\theta).$$

Using $\mathcal{L} = T - V$, $q = \theta$ and $\dot{q} = \dot{\theta}$, we have

$$\frac{\partial \mathcal{L}}{\partial \theta} - \frac{d}{dt}\left(\frac{\partial \mathcal{L}}{\partial \dot{\theta}}\right) = 0,$$

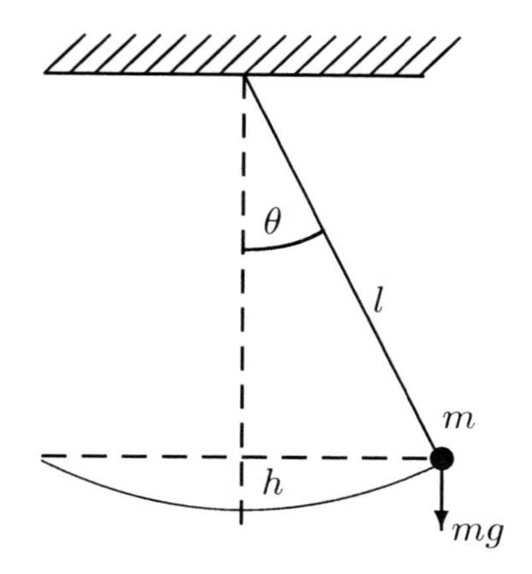

Figure 5.2: A simple pendulum.

which becomes

$$-mgl\sin\theta - \frac{d}{dt}(ml^2\dot{\theta}) = 0.$$

Therefore, we have the pendulum equation

$$\frac{d^2\theta}{dt^2} + \frac{g}{l}\sin\theta = 0.$$

This is a nonlinear equation. If the angle is very small ($\theta \ll 1$), $\sin\theta \approx \theta$, we then have the standard equation for the linear harmonic motion

$$\frac{d^2\theta}{dt^2} + \frac{g}{l}\theta = 0.$$

◇

5.2 Variations with Constraints

Although the stationary requirement in the calculus of variations leads to the minimization of the integral itself, there is no additional constraint. In this sense, the calculus of variation discussed up to now is unconstrained. However, sometimes these variations have certain additional constraints, for example, the sliding of a bead on a hanging string. Now we want to make the integral I stationary under another constraint integral Q that is constant. We have

$$I = \int_a^b \psi(x, y, y')dx, \tag{5.24}$$

subjected to the constraint

$$Q = \int_a^b \phi(x, y, y')dx. \tag{5.25}$$

As for most optimization problems under additional constraints, the method of Lagrange multipliers can transform the constrained problem into an unconstrained one by using a combined functional $J = I + \lambda Q$ or

$$J = \int_a^b [\psi + \lambda\phi]dx, \tag{5.26}$$

where λ is the undetermined Lagrange multiplier. Replacing ψ by $[\psi + \lambda\phi]$ in the Euler-Lagrange equation or following the same procedure of the derivations, we have

$$[\frac{\partial\psi}{\partial y} - \frac{d}{dx}(\frac{\partial\psi}{\partial y'})] + \lambda[\frac{\partial\phi}{\partial y} - \frac{d}{dx}(\frac{\partial\phi}{\partial y'})] = 0. \tag{5.27}$$

Now we can come back to our example of the hanging rope problem with two fixed points. The total length of the rope is L, and it hangs from two fixed points $(-d, 0)$ and $(d, 0)$. From the geometric consideration, it requires that $2d < L$. In order to find the shape of the hanging rope under gravity, we now define its gravitational potential energy E_p as

$$E_p = \int_{-d}^{d} [\rho g y(x) ds] = \rho g \int_{-d}^{d} y\sqrt{1 + y'^2}dx. \tag{5.28}$$

The additional constraint is that the total length of the rope is a constant (L). Thus,

$$Q = \int_{-d}^{d} \sqrt{1 + y'^2}dx = L. \tag{5.29}$$

By using the Lagrange multiplier λ, we have $J = E_p + \lambda Q$, or

$$J = \int_{-d}^{d} [\rho g y + \lambda] \sqrt{1 + y'^2} dx. \quad (5.30)$$

Since $\Psi = [\rho g y + \lambda]\sqrt{1 + y'^2}$ does not contain x explicitly, or $\frac{\partial \Psi}{\partial x} = 0$, then the Euler-Lagrange equation can be reduced into a simpler form in this special case. Using

$$\frac{d\Psi}{dx} = \frac{\partial \Psi}{\partial x} + \frac{\partial \Psi}{\partial y}\frac{dy}{dx} + \frac{\partial \Psi}{\partial y'}\frac{dy'}{dx}$$

$$= 0 + y'\frac{\partial \Psi}{\partial y} + y''\frac{\partial \Psi}{\partial y'}, \quad (5.31)$$

and the Euler-Lagrange equation $\frac{\partial \Psi}{\partial y} = \frac{d}{dx}(\frac{\partial \Psi}{\partial y'})$, we have

$$\frac{d\Psi}{dx} = y'[\frac{d}{dx}(\frac{\partial \Psi}{\partial y'})] + y''\frac{\partial \Psi}{\partial y'} = \frac{d}{dx}[y'\frac{\partial \Psi}{\partial y'}], \quad (5.32)$$

which can again be written as

$$\frac{d}{dx}[\Psi - y'\frac{\partial \Psi}{\partial y'}] = 0. \quad (5.33)$$

The integration of this equation gives

$$\Psi - y'\frac{\partial \Psi}{\partial y'} = A = const. \quad (5.34)$$

Substituting the expression of Ψ into the above equation, the stationary values of J requires

$$\sqrt{1 + y'^2} - \frac{y'^2}{\sqrt{1 + y'^2}} = \frac{A}{\rho g y + \lambda}. \quad (5.35)$$

Multiplying both sides by $\sqrt{1 + y'^2}$ and using the substitution $A \cosh \zeta = \rho g y + \lambda$, we have

$$y'^2 = \cosh^2 \zeta - 1, \quad (5.36)$$

whose solution is

$$\cosh^{-1}[\frac{\rho g y + \lambda}{A}] = \frac{x \rho g}{A} + K. \tag{5.37}$$

Using the boundary conditions at $x = \pm d$ and the constraint $Q = L$, we have $K = 0$ and implicit equation for A

$$\sinh(\frac{\rho g d}{A}) = \frac{\rho g L}{2A}. \tag{5.38}$$

Finally, the curve for the hanging rope becomes the following catenary

$$y(x) = \frac{A}{\rho g}[\cosh(\frac{\rho g x}{A}) - \cosh(\frac{\rho g d}{A})]. \tag{5.39}$$

◇ **Example 5.3:** For the hanging rope problem, what happens if we only fix one end at $(a, 0)$, while allowing the free end of the hanging rope to slide on a vertical pole? Well, this forms a variation problem with variable end-point(s). We assume that free end is at $(0, y)$ where y acts like a free parameter to be determined. Now the boundary condition at the free end is different. Since the variation of $\delta I = 0$, we have

$$\delta J = \int_a^b [\frac{\partial \Psi}{\partial y} - \frac{d}{dx}(\frac{\partial \Psi}{\partial y'})]\zeta dx + [\zeta \frac{\partial \Psi}{\partial y'}]_a^b = 0.$$

As the variation ζ is now non-zero at the free end point, we then have

$$\frac{\partial \Psi}{\partial y'} = 0.$$

From $J = E_p + \lambda Q$, we have $\Psi = (\rho g y + \lambda)\sqrt{1 + y'^2}$. Thus, we get

$$\frac{\partial}{\partial y'}[(\rho g y + \lambda)\sqrt{1 + y'^2}] = 0,$$

or

$$y'(\rho g y + \lambda)/\sqrt{1 + y'^2} = 0, \qquad \text{or} \qquad y' = 0.$$

In other words, the slope is zero at the free end. ◇

Such a boundary condition of $y' = 0$ has the real physical meaning because any non-zero gradient at the free end would have a non-zero vertical component, thus causing the vertical slip along the rope due to the tension in the rope. The zero-gradient leads to the static equilibrium. Thus, the whole curve of the hanging rope with one free end forms half the catenary.

◇ **Example 5.4:** Dido's problem concerns the strategy to enclose a maximum area with a fixed length circumference. Legend says that Dido was promised a piece of land on the condition that it was enclosed by an oxhide. She had to cover as much as land as possible using the given oxhide. She cut the oxhide into narrow strips with ends joined, and a whole region of a hill was enclosed.

Suppose the total length of the oxhide strip is L. The enclosed area A to be maximized is

$$A = \int_{x_a}^{x_b} y(x)dx,$$

where x_a and x_b are two end points (of course they can be the same points). We also have the additional constraint

$$\int_{x_a}^{x_b} \sqrt{1+y'^2}dx = L = const.$$

This forms an isoperimetric variation problem. As L is fixed, thus the maximization of A is equivalent to make $I = A + \lambda L$ stationary. That is

$$I = A + \lambda L = \int_{x_a}^{x_b} [y + \lambda\sqrt{1+y'^2}]dx.$$

Using the Euler-Lagrange equation, we have

$$\frac{\partial I}{\partial y} - \frac{d}{dx}\frac{\partial I}{\partial y'} = 0,$$

or

$$\frac{\partial}{\partial y}[y + \lambda\sqrt{1+y'^2}] + \frac{d}{dx}\frac{\partial}{\partial y'}[y + \lambda\sqrt{1+y'^2}] = 0,$$

which becomes

$$1 - \lambda\frac{d}{dx}\Big(\frac{y'}{\sqrt{1+y'^2}}\Big) = 0.$$

Integrating it once, we get

$$\frac{\lambda y'}{\sqrt{1+y'^2}} = x + K,$$

where K is the integration constant. By rearranging, we have

$$y' = \pm\frac{x+K}{\sqrt{\lambda^2 - (x+K)^2}}.$$

Integrating this equation again, we get

$$y(x) = \mp\sqrt{\lambda^2 - (x+K)^2} + B,$$

where B is another integration constant. This is equivalent to

$$(x+K)^2 + (y-B)^2 = \lambda^2,$$

which is essentially the standard equation for a circle with the centre at $(-K, B)$ and a radius λ. Therefore, the most area that can be enclosed by a fixed length is a circle. ◇

An interesting application is the design of the slides in playgrounds. Suppose we want to design a smooth (frictionless) slide, what is the best curve/shape the slide should take so that a child can slide down in a quickest way? This problem is related to the brachistochrone problem, also called the shortest time problem or steepest descent problem, which initiated the development of the calculus of variations. In 1696, Johann Bernoulli posed a problem to find the curve that minimizes the time for a bead attached to a wire to slide from a point $(0, h)$ to a lower point $(a, 0)$. It was believed that Newton solved it within a few hours after receiving it. From the conservation of energy, we can determine the speed of the bead from the equation $\frac{1}{2}mv^2 + mgy = mgh$, and we have

$$v = \sqrt{2g(h-y)}. \tag{5.40}$$

So the total time taken to travel from $(0, h)$ to $(a, 0)$ is

$$t = \int_0^a \frac{1}{v} ds = \int_0^a \frac{\sqrt{1+y'^2}}{\sqrt{2g(h-y)}} dx. \tag{5.41}$$

Using the simplified Euler-Lagrange equation (5.34) because the integrand $\Psi = \sqrt{1+y'^2}/\sqrt{2g(h-y)}$ does not contain x explicitly, we have

$$\sqrt{\frac{(1+y'^2)}{2g(h-y)}} - y' \frac{\partial}{\partial y'}[\sqrt{\frac{(1+y'^2)}{2g(h-y)}}] = A. \tag{5.42}$$

By differentiation and some rearrangements, we have

$$y'^2 = \frac{B-h+y}{h-y}, \qquad B = \frac{1}{2gA^2}. \tag{5.43}$$

By changing of variables $\eta = h - y = \frac{B}{2}(1 - \cos\theta)$ and integrating, we have

$$x = \frac{B}{2}[\theta - \sin\theta] + k, \tag{5.44}$$

where $\theta < \pi$ and k is an integration constant. As the curve must pass the point $(0, h)$, we get $k = 0$. So the parametric equations for the curve become

$$x = \frac{B}{2}(\theta - \sin\theta), \qquad y = h - \frac{B}{2}(1 - \cos\theta). \tag{5.45}$$

This is a cycloid, not a straight line, which seems a bit surprising, or at least it is rather counter-intuitive. The bead travels a longer distance, thus has a higher average velocity and subsequently falls quicker than traveling in a straight line.

5.3 Variations for Multiple Variables

What we have discussed so far mainly concerns the variations in 2-D, and subsequently the variations are in terms $y(x)$ or curves only. What happens if we want to study a surface in the full 3-D configuration? The principle in the previous sections can be extended to any dimensions with multiple variables, however, we will focus on the minimization of a surface here. Suppose we want to study the shape of a soap bubble, the principle of least action leads to the minimal surface problem. The surface integral of a soap bubble should be stationary. Now we assume that the shape of the bubble is $u(x, y)$, then the total surface area is

$$A(u) = \iint_\Omega \Psi dxdy = \iint_\Omega \sqrt{1 + (\frac{\partial u}{\partial x})^2 + (\frac{\partial u}{\partial y})^2} dxdy, \tag{5.46}$$

where

$$\Psi = \sqrt{1 + (\frac{\partial u}{\partial x})^2 + (\frac{\partial u}{\partial y})^2} = \sqrt{1 + u_x^2 + u_y^2}. \tag{5.47}$$

In this case, the extended Euler-Lagrangian equation for two variables x and y becomes

$$\frac{\partial \Psi}{\partial u} - \frac{\partial}{\partial x}(\frac{\partial \Psi}{\partial u_x}) - \frac{\partial}{\partial y}(\frac{\partial \Psi}{\partial u_y}) = 0. \tag{5.48}$$

Substituting Ψ into the above equation and using $\frac{\partial \Psi}{\partial u} = \Psi_u = 0$ since Ψ does not contain u explicitly, we get

$$-\frac{\partial}{\partial x}[\frac{1}{\Psi}\frac{\partial u}{\partial x}] - \frac{\partial}{\partial y}[\frac{1}{\Psi}\frac{\partial u}{\partial y}] = 0, \tag{5.49}$$

or

$$(1 + u_y^2)u_{xx} - 2u_x u_y + (1 + u_x^2)u_{yy} = 0. \tag{5.50}$$

This is a nonlinear equation and its solution is out of the scope of this book. This nonlinear equation has been one of the active research topics for more than a century. It has been proved that the fundamental solution to this equation is a sphere, and in fact we know that all bubbles are spherical. For some problems, we can approximately assume that u_x and u_y are small, thus the above equation becomes Laplace's equation

$$u_{xx} + u_{yy} = 0. \tag{5.51}$$

The calculus of variations has many applications. The other classical examples include Fermat's principle in optics, Sturm-Liouville problem, surface shape minimization, the action principle, and of course the finite element analysis.

5.4 Oscillations and Modes

Vibrations and oscillations forms an important topic in finite element analysis and the mode analysis. It shares a great similarity between oscillation components and finite elements. The simplest form is the harmonic motion of a pendulum as shown in Fig. 5.2. In this case, we know its period T and thus its natural frequency $\omega_0 = 2\pi/T$. In general, a system may have many natural frequencies or modes in a system, and the natural frequencies are in fact determined from the eigenvalue problem resulting from the system.

5.4.1 Normal Modes

Now let us study a more complicated system with three mass blocks attached in by two springs as shown in Figure 5.3. This system can be thought of as a car attached to two caravans on a flat road. Let u_1, u_2, u_3 be the displacement of the three mass blocks m_1, m_2, m_3, respectively. Then, their accelerations will be $\ddot{u}_1$, $\ddot{u}_2$, $\ddot{u}_3$ where $\ddot{u} = d^2u/dt^2$. From the balance of forces and Newton's law, we have

$$m_1\ddot{u}_1 = k_1(u_2 - u_1), \tag{5.52}$$

$$m_2\ddot{u}_2 = k_2(u_3 - u_2) - k_1(u_2 - u_1), \tag{5.53}$$

$$m_3\ddot{u}_3 = -k_2(u_3 - u_2). \tag{5.54}$$

These equations can be written in a matrix form as

$$\begin{pmatrix} m_1 & 0 & 0 \\ 0 & m_2 & 0 \\ 0 & 0 & m_3 \end{pmatrix} \begin{pmatrix} \ddot{u}_1 \\ \ddot{u}_2 \\ \ddot{u}_3 \end{pmatrix}$$

$$+ \begin{pmatrix} k_1 & -k_1 & 0 \\ -k_1 & k_1 + k_2 & -k_2 \\ 0 & -k_2 & k_2 \end{pmatrix} \begin{pmatrix} u_1 \\ u_2 \\ u_3 \end{pmatrix} = \begin{pmatrix} 0 \\ 0 \\ 0 \end{pmatrix}, \tag{5.55}$$

or

$$\mathbf{M}\ddot{\mathbf{u}} + \mathbf{K}\mathbf{u} = \mathbf{0}, \tag{5.56}$$

where $\mathbf{u}^T = (u_1, u_2, u_3)$. The mass matrix is

$$\mathbf{M} = \begin{pmatrix} m_1 & 0 & 0 \\ 0 & m_2 & 0 \\ 0 & 0 & m_3 \end{pmatrix}, \tag{5.57}$$

and the stiffness matrix is

$$\mathbf{K} = \begin{pmatrix} k_1 & -k_1 & 0 \\ -k_1 & k_1 + k_2 & -k_2 \\ 0 & -k_2 & k_2 \end{pmatrix}. \tag{5.58}$$

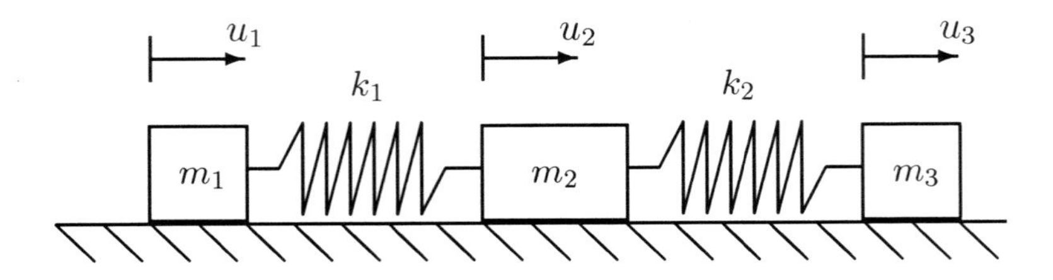

Figure 5.3: Harmonic vibrations.

Equation (5.56) is a second-order ordinary differential equation in terms of matrices. This homogeneous equation can be solved by substituting $u_i = U_i \cos(\omega t)$ where $U_i (i = 1, 2, 3)$ are constants and ω^2 can have several values which correspond to the natural frequencies. Now we have

$$-\omega_i^2 \mathbf{M} U_i \cos(\omega t) + \mathbf{K} U_i \cos(\omega t) = 0, \tag{5.59}$$

where $i = 1, 2, 3$. Dividing both sides by $\cos(\omega t)$, we have

$$[\mathbf{K} - \omega^2 \mathbf{M}] U_i = 0. \tag{5.60}$$

This is essentially an eigenvalue problem because any non-trivial solutions for U_i require

$$|\mathbf{K} - \omega^2 \mathbf{M}| = 0. \tag{5.61}$$

Therefore, the eigenvalues of this equation give the natural frequencies.

◇ **Example 5.5:** For the simplest case when $m_1 = m_2 = m_3 = m$ and $k_1 = k_2 = k$, we have

$$\begin{vmatrix} k - \omega^2 m & -k & 0 \\ -k & 2k - \omega^2 m & -k \\ 0 & -k & k - \omega^2 m \end{vmatrix} = 0,$$

or

$$-\omega^2(k-\omega^2 m)(3km-\omega^2 m^2)=0.$$

This is a cubic equation in terms of ω^2, and it has three solutions. Therefore, the three natural frequencies are

$$\omega_1^2=0, \qquad \omega_2^2=\frac{k}{m}, \qquad \omega_3^2=\frac{3k}{m}.$$

For $\omega_1^2=0$, we have $(U_1,U_2,U_3)=\frac{1}{\sqrt{3}}(1,1,1)$, which is the rigid body motion. For $\omega_2=k/m$, the eigenvector is determined by

$$\begin{pmatrix} 0 & -k & 0 \\ -k & k & -k \\ 0 & -k & 0 \end{pmatrix}\begin{pmatrix} U_1 \\ U_2 \\ U_3 \end{pmatrix}=\begin{pmatrix} 0 \\ 0 \\ 0 \end{pmatrix},$$

which leads to $U_2=0$, and $U_1=U_3$. Written in normalized form, it becomes $(U_1,U_2,U_3)=\frac{1}{\sqrt{2}}(1,0,-1)$. This means that block 1 moves in the opposite direction away from block 3, and block 2 remains stationary. For $\omega_3^2=3k/m$, we have $(U_1,U_2,U_3)=\frac{1}{\sqrt{6}}(1,-2,1)$. That is to say, block 2 moves in the different direction from block 3 which is at the same pace with block 1. ◇

5.4.2 Small Amplitude Oscillations

For a mechanically conservative system, its total energy $E=T+V$ is conserved, where T is its total kinetic energy and V is its total potential energy. The configuration of the mechanical system can be described by its general coordinates $\mathbf{q}=(q_1,q_2,...,q_n)$. The general coordinates can be distance and angles. Thus, the velocities of the system will be $\dot{\mathbf{q}}=\dot{q}_1,\dot{q}_2,...,\dot{q}_n$. If we consider the system consists of many small particles or even imaginary parts, then the total kinetic energy T is a function of $\dot{\mathbf{q}}$ and sometimes $\mathbf{q}$, but the potential energy V is mainly a function of $\mathbf{q}$ only. As we are only concerned with small amplitude oscillations near equilibrium $V_{\min}=V(0)=V_0$, we can always take $\mathbf{q}=0$ at the equilibrium so that we can expand V in terms of $\mathbf{q}$ as a Taylor series

$$V(\mathbf{q})=V_{min}+\sum_i \frac{\partial V_0}{\partial q_i}q_i+\sum_i\sum_j K_{ij}q_iq_j+..., \tag{5.62}$$

where the stiffness matrix is

$$K_{ij} = \frac{1}{2} \frac{\partial^2 V_0}{\partial q_i \partial q_j}\Big|_{q_i=0, q_j=0}. \qquad (5.63)$$

Since potential energy is always relative to an arbitrary reference point, we can thus take the potential energy at equilibrium V_{min} to be zero. In addition, the equilibrium or the minimum value of V requires $\frac{\partial V}{\partial q_i} = 0$ at the equilibrium point $q_i = 0$, and the force $F_i = \frac{\partial V}{\partial q_i}$ shall be zero. This is correct because the resultant force must be zero at equilibrium, otherwise, the system will be driven away by the resultant force. The component of the resultant force along the general coordinate q_i should also be zero. Therefore, the total potential energy is now simplified as

$$V = \sum_i \sum_j q_i K_{ij} q_j = \mathbf{q}^T \mathbf{K} \mathbf{q}, \qquad (5.64)$$

which is a quadratic form.

For any small oscillation, the velocity is linear in terms of $\dot{q}_i$, and thus the corresponding kinetic energy is $\frac{1}{2} m \dot{q}_i^2$. The total kinetic energy is the sum of all the components over all particles or parts, forming a quadratic form. That is to say,

$$T = \sum_i \sum_j m_{ij} \dot{q}_i \dot{q}_j = \dot{\mathbf{q}}^T \mathbf{M} \dot{\mathbf{q}}, \qquad (5.65)$$

where $\mathbf{M} = [m_{ij}]$ is the mass matrix.

For a conservative system, the total mechanical energy $E = T + V$ is conserved, and thus time-independent. So we have

$$\frac{d(T+V)}{dt} = \frac{d}{dt}[\dot{\mathbf{q}}^T M \dot{\mathbf{q}} + \mathbf{q}^T \mathbf{K} \mathbf{q}] = 0. \qquad (5.66)$$

Since $\mathbf{M}$ and $\mathbf{K}$ are symmetric matrices, this above equation becomes

$$\mathbf{M}\ddot{\mathbf{q}} + \mathbf{K}\mathbf{q} = 0. \qquad (5.67)$$

This is a second order ordinary differential equation for matrices. Assuming the solution in the form $\mathbf{q}^T = (q_1, q_2, ..., q_n) =$

$(U_1 \cos \omega t, U_2 \cos \omega t, ..., U_n \cos \omega t)$ and substituting it into the above equation, we have

$$|\mathbf{M} - \omega^2 \mathbf{K}| = 0, \tag{5.68}$$

which is an eigenvalue problem.

As an application, let us solve the same system of three mass blocks discussed earlier as shown in Figure 5.3. The total potential energy T is the sum of each mass block

$$T = \frac{1}{2} m_1 (\dot{u}_1)^2 + \frac{1}{2} m_2 (\dot{u}_2)^2 + \frac{1}{2} m_3 (\dot{u}_3)^2$$

$$= \begin{pmatrix} \dot{u}_1 & \dot{u}_2 & \dot{u}_3 \end{pmatrix} \begin{pmatrix} m_1 & 0 & 0 \\ 0 & m_2 & 0 \\ 0 & 0 & m_3 \end{pmatrix} \begin{pmatrix} \dot{u}_1 \\ \dot{u}_2 \\ \dot{u}_3 \end{pmatrix}, \tag{5.69}$$

which can be written as a quadratic form

$$T = \dot{\mathbf{u}}^T \mathbf{M} \dot{\mathbf{u}}, \tag{5.70}$$

where $\mathbf{u}^T = (u_1, u_2, u_3)$, and

$$\mathbf{M} = \frac{1}{2} \begin{pmatrix} m_1 & 0 & 0 \\ 0 & m_2 & 0 \\ 0 & 0 & m_3 \end{pmatrix}. \tag{5.71}$$

We see that $\mathbf{M}$ is a symmetric matrix.

For a spring, the force is $f = kx$, thus the potential energy stored in a spring is

$$\int_0^u kx dx = \frac{1}{2} k u^2. \tag{5.72}$$

Therefore, the total potential energy of the two-spring system is

$$V = \frac{1}{2} k_1 (u_2 - u_1)^2 + \frac{1}{2} k_2 (u_3 - u_2)^2. \tag{5.73}$$

Since interchange of u_1 and u_2 does not change V, it is thus symmetric in terms of u_1, u_2 etc, which implies that K_{ij} should be symmetric as well.

The stiffness matrix $\mathbf{K} = [K_{ij}]$ can be calculated using

$$K_{ij} = \frac{1}{2} \frac{\partial^2 V}{\partial u_i \partial u_j}. \tag{5.74}$$

For example,

$$K_{11} = \frac{1}{2} \frac{\partial^2 V}{\partial {u_1}^2} = \frac{1}{2} \times k_1 = \frac{k_1}{2}, \tag{5.75}$$

and

$$K_{12} = \frac{1}{2} \frac{\partial^2 V}{\partial u_1 \partial u_2} = \frac{1}{2} \frac{\partial}{\partial u_1} (\frac{\partial V}{\partial u_2})$$

$$= \frac{1}{2} \frac{\partial}{\partial u_1} [k_1(u_2 - u_1) + k_2(u_3 - u_2)] = -\frac{k_1}{2}. \tag{5.76}$$

Following the similar calculations, we have

$$\mathbf{K} = \frac{1}{2} \begin{pmatrix} k_1 & -k_1 & 0 \\ -k_1 & k_1 + k_2 & k_2 \\ 0 & -k_2 & k_2 \end{pmatrix}, \tag{5.77}$$

which is exactly 1/2 multiplying the stiffness matrix we obtained earlier in equation (5.58). Thus, the equation for small amplitude oscillation is

$$\mathbf{M}\ddot{\mathbf{u}} + \mathbf{K}\mathbf{u} = 0. \tag{5.78}$$

For the special case of $m_1 = m_2 = m_3 = m$ and $k_1 = k_2 = k$, its eigenvalues are determined by

$$\begin{vmatrix} k - \omega^2 m & -k & 0 \\ -k & 2k - \omega^2 m & -k \\ 0 & -k & k - \omega^2 m \end{vmatrix} = 0, \tag{5.79}$$

which is exactly the problem we solved in the previous section (see example 5.5).

For a simple system such as a pendulum, equation (5.58) is equivalent to the following simple formula for calculating the natural frequency

$$\omega = \sqrt{\frac{V''(q)}{M(q)}}, \tag{5.80}$$

where $V'' > 0$ because the potential energy at equilibrium is minimum.

◇ **Example 5.6:** A simple pendulum with a mass m is hanged vertically from a ceiling with a distance L from the fixed point. Let θ be the small angle from its equilibrium, then the kinetic energy is $T = \frac{1}{2}mv^2 = \frac{1}{2}mL^2(\dot{\theta})^2$. The potential energy is

$$V = mgL(1 - \cos\theta).$$

Therefore, the stiffness is $K = \frac{1}{2}V''(\theta) = \frac{1}{2}mgL\cos\theta|_{\theta=0} = mgL/2$. The equivalent mass is $M(\theta) = \frac{1}{2}mL^2$. The governing equation becomes

$$\frac{1}{2}mL^2\ddot{\theta} + \frac{1}{2}mgL\theta,$$

or

$$\ddot{\theta} + \frac{L}{g}\theta = 0.$$

The natural frequency for small oscillations is

$$\omega = \sqrt{\frac{V''}{M(q)}} = \sqrt{\frac{g}{L}}.$$

The period of this pendulum is

$$\tau = \frac{2\pi}{\omega} = 2\pi\sqrt{\frac{L}{g}}.$$

--- ◇

Chapter 6

Finite Element Method

In the finite difference methods, we approximate the equations at a finite number of discrete points, and there are many limitations in finite difference methods. One of such disadvantages is that it is not straightforward to deal with irregular geometry. More versatile and efficient methods are highly needed. In fact, the finite element method is one class of the most successful methods in scientific computing and has a wide range of applications.

The basic idea of finite element analysis is to divide the domain into main small blocks or elements as shown in Fig. 6.1. This is equivalent to imaginarily cut a solid structure such as a building or bridge into many pieces or elements. These small blocks are characterised by nodes shown as solid dots in Fig. 6.1, and the whole domain can be considered as if these blocks or elements are glued together at these nodes and along the element boundaries. In this way, we essentially transform a continuum system with infinite degrees of freedom into a discrete finite system with finite degrees of freedom. In fact, this is the origin of the name of finite elements. As we know that most continuum systems are governed by differential equations, the major advantage of this transformation is that the differential equation for a continuum system is transformed into a set of simultaneous algebraic equations for the discrete system with finite number of elements. The approximations to any field quantities such displacements and stresses over these finite elements use the piecewise polynomial interpolation techniques.

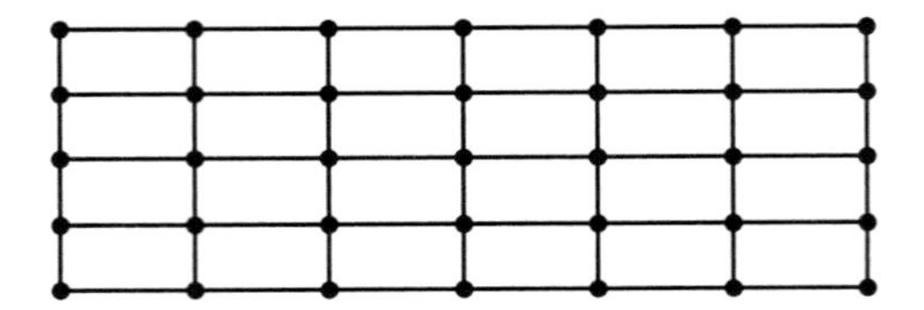

Figure 6.1: Concept of finite elements.

The fundamental aim of the finite element analysis is to formulate the numerical method in such a way that the partial differential equation in combination with the appropriate boundary conditions and loads will be transformed into algebraic equations in terms of matrices. For time-dependent problems involving partial differential equations, the equations will be transformed into an ordinary differential equation in terms of matrices, which will in turn be discretized and converted into algebraic equations by time-stepping or some iteration techniques. For example, a linear elastic problem can be formulated in such a way that it is equivalent to the equation of the following type

$$\mathbf{K}\mathbf{u} = \mathbf{f}, \tag{6.1}$$

where $\mathbf{K}$ is the stiffness matrix, and $\mathbf{f}$ is a vector corresponding to nodal forces and some contribution from boundary conditions. $\mathbf{u}$ is the unknown vector to be solved and it corresponds to the nodal degree of freedom such as the displacement.

6.1 Concept of Elements

6.1.1 Simple Spring Systems

The basic idea of the finite element analysis is to divide a model (such as a bridge and an airplane) into many pieces or elements with discrete nodes. These elements form an approximate system to the whole structures in the domain of interest, so that the physical quantities such as displacements can be evaluated at these discrete nodes. Other quantities such as stresses, strains can then be evaluated at certain points (usually Gaussian integration points) inside elements. The simplest elements are the element with two nodes in 1-D, the triangular element with three nodes in 2-D, and tetrahedral elements with four nodes in 3-D.

In order to show the basic concept, we now focus on the simplest 1-D spring element with two nodes (see Figure 6.2). The spring has a stiffness constant k (N/m) with two nodes i and j. At nodes i and j, the displacements (in metres) are u_i and u_j, respectively. f_i and f_j are nodal forces.

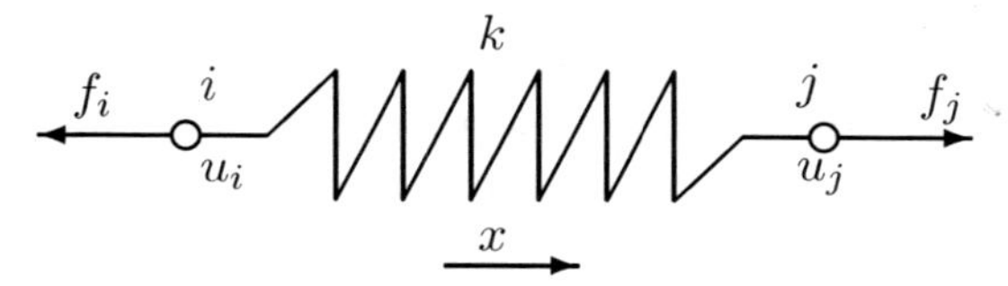

Figure 6.2: A spring element.

From Hooke's law, we know the displacement $\Delta u = u_j - u_i$ is related to f, or

$$f = k(\Delta u). \tag{6.2}$$

At node i, we have

$$f_i = -f = -k(u_j - u_i) = ku_i - ku_j, \tag{6.3}$$

and at node j, we get

$$f_j = f = k(u_j - u_i) = -ku_i + ku_j. \tag{6.4}$$

These two equations can be combined into a matrix equation

$$\begin{pmatrix} k & -k \\ -k & k \end{pmatrix} \begin{pmatrix} u_i \\ u_j \end{pmatrix} = \begin{pmatrix} f_i \\ f_j \end{pmatrix}, \qquad \text{or} \qquad \mathbf{K}\mathbf{u} = \mathbf{f}. \tag{6.5}$$

Here $\mathbf{K}$ is the stiffness matrix, $\mathbf{u}$ and $\mathbf{f}$ are the displacement vector and force vector, respectively. This is the basic spring element, and let us see how it works in a spring system such as

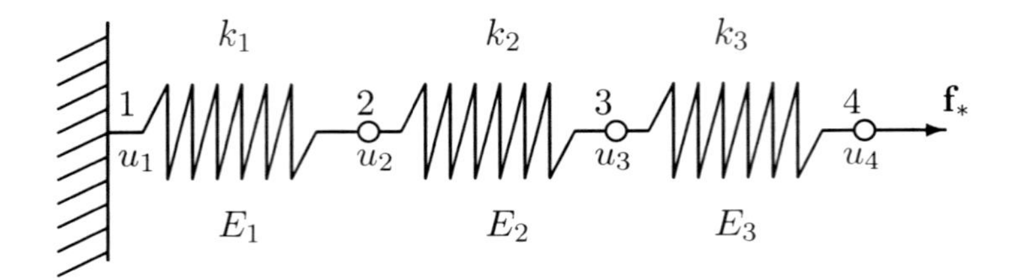

Figure 6.3: A simple spring system.

shown in Figure 6.3 where three different springs are connected in series.

For a simple spring system shown in Figure 6.3, we now try to determine the displacements of $u_i (i = 1, 2, 3, 4)$. In order to do so, we have to assemble the whole system into a single equation in terms of global stiffness matrix $\mathbf{K}$ and forcing $\boldsymbol{f}$. As these three elements are connected in series, the assembly of the system can be done element by element. For element E_1, its contribution to the overall global matrix is

$$\begin{pmatrix} k_1 & -k_1 \\ -k_1 & k_1 \end{pmatrix} \begin{pmatrix} u_1 \\ u_2 \end{pmatrix} = \begin{pmatrix} f_1 \\ f_2 \end{pmatrix}, \tag{6.6}$$

which is equivalent to

$$\mathbf{K}_1 \mathbf{u} = \mathbf{f}_{E_1}, \tag{6.7}$$

where

$$\begin{pmatrix} k_1 & -k_1 & 0 & 0 \\ -k_1 & k_1 & 0 & 0 \\ 0 & 0 & 0 & 0 \\ 0 & 0 & 0 & 0 \end{pmatrix} \begin{pmatrix} u_1 \\ u_2 \\ u_3 \\ u_4 \end{pmatrix} = \begin{pmatrix} f_1 \\ f_2 \\ . \\ . \end{pmatrix}, \tag{6.8}$$

and $\mathbf{f}_{E_1}^T = (f_1, f_2, 0, 0)$. Similarly, for element E_2, we have

$$\begin{pmatrix} k_2 & -k_2 \\ -k_2 & k_2 \end{pmatrix} \begin{pmatrix} u_2 \\ u_3 \end{pmatrix} = \begin{pmatrix} -f_2 \\ f_3 \end{pmatrix}, \tag{6.9}$$

or

$$\mathbf{K_2} = \begin{pmatrix} 0 & 0 & 0 & 0 \\ 0 & k_2 & -k_2 & 0 \\ 0 & -k_2 & k_2 & 0 \\ 0 & 0 & 0 & 0 \end{pmatrix}, \tag{6.10}$$

where we have used the balance at node 2. For element E_3, we have

$$\begin{pmatrix} k_3 & -k_3 \\ -k_3 & k_3 \end{pmatrix} \begin{pmatrix} u_3 \\ u_4 \end{pmatrix} = \begin{pmatrix} -f_3 \\ f_* \end{pmatrix}, \tag{6.11}$$

or

$$\mathbf{K}_3 = \begin{pmatrix} 0 & 0 & 0 & 0 \\ 0 & 0 & 0 & 0 \\ 0 & 0 & k_3 & -k_3 \\ 0 & 0 & -k_3 & k_3 \end{pmatrix}, \tag{6.12}$$

where $f_4 = f_*$ has been used. We can now add the three sets of equations together to obtain a single equation

$$\begin{pmatrix} k_1 & -k_2 & 0 & 0 \\ -k_1 & k_1 + k_2 & -k_2 & 0 \\ 0 & -k_2 & k_2 + k_3 & -k_3 \\ 0 & 0 & -k_3 & k_3 \end{pmatrix} \begin{pmatrix} u_1 \\ u_2 \\ u_3 \\ u_4 \end{pmatrix} = \begin{pmatrix} f_1 \\ -f_2 + f_2 \\ -f_3 + f_3 \\ f_* \end{pmatrix},$$

or

$$\mathbf{Ku} = \mathbf{f}, \tag{6.13}$$

where

$$\mathbf{K} = \mathbf{K}_1 + \mathbf{K}_2 + \mathbf{K}_3$$

$$= \begin{pmatrix} k_1 & -k_1 & 0 & 0 \\ -k_1 & k_1 + k_2 & -k_2 & 0 \\ 0 & -k_2 & k_2 + k_3 & -k_3 \\ 0 & 0 & -k_3 & k_3 \end{pmatrix}, \tag{6.14}$$

and

$$\mathbf{u}^T = (u_1, u_2, u_3, u_4), \qquad \mathbf{f} = \mathbf{f}_{E_1} + \mathbf{f}_{E_2} + \mathbf{f}_{E_3}. \tag{6.15}$$

In general, the matrix $\mathbf{K}$ is singular or its rank is less than the total number of degrees of freedom, which is four in this case. This means that the equation has no unique solution. Thus, we need the boundary conditions to ensure a unique solution. In this spring system, if no boundary condition is applied at any nodes, then the applied force at the node 4 will make the spring system fly to the right. If we add a constraint by fixing the left node 1, then the system can stretch, and a unique configuration is formed.

In our case where there are no applied forces at nodes 2 and 3, we have

$$\mathbf{f}^T = (0, 0, 0, f_*). \tag{6.16}$$

◇ **Example 6.1:** For $k_1 = 100$ N/m, $k_2 = 200$ N/m, and $k_3 = 50$N/m, and $f_* = 20$ N, the boundary at node 1 is fixed ($u_1 = 0$). Then, the stiffness matrix is

$$\mathbf{K} = \begin{pmatrix} 100 & -100 & 0 & 0 \\ -100 & 300 & -200 & 0 \\ 0 & -200 & 250 & -50 \\ 0 & 0 & -50 & 50 \end{pmatrix},$$

and the force column vector

$$\mathbf{f}^T = (0, 0, 0, 20).$$

The rank of $\mathbf{K}$ is 3, therefore, we need at least one boundary condition. By applying $u_1 = 0$, we now have only three unknown displacements u_2, u_3, u_4. Since $u_1 = 0$ is already known, the first equation for u_1 becomes redundant and we can now delete it so that the reduced stiffness matrix $\mathbf{A}$ is a 3×3 matrix. Therefore, we have

$$\mathbf{A} = \begin{pmatrix} 300 & -200 & 0 \\ -200 & 250 & 0 \\ 0 & -50 & 50 \end{pmatrix},$$

and the reduced forcing vector is

$$\mathbf{g}^T = (0, 0, 20).$$

The solution is

$$\mathbf{u} = \mathbf{A}^{-1}\mathbf{g} = \begin{pmatrix} 0.2 \\ 0.3 \\ 0.7 \end{pmatrix}.$$

Therefore, the displacements are $u_2 = 0.2$ m, $u_3 = 0.3$ m, and $u_4 = 0.7$ m.

Theoretically speaking, the force should be 20N everywhere in the spring systems since the mass of the springs is negligible. Let us calculate the force at nodes 2 and 3 to see if this is the case. At the node 2, the extension in element E_1 is $\Delta u = u_2 - u_1 = 0.2$ m, thus the force at node 2 is

$$f_2 = k_1 \Delta u = 100 \times 0.2 = 20\text{N}.$$

Similarly, at node 3 of element E_2, we have

$$f_3 = k_2(u_3 - u_2) = 200 \times 0.1 = 20\text{N},$$

which is the same at node 3 of element E_3

$$f_3 = k_3 \times (-\Delta u) = k_3(u_4 - u_3) = 50 \times 0.4 = 20\text{N}.$$

So the force is 20 N everywhere. ◇

6.1.2 Bar and Beam Elements

The spring system we discussed earlier is limited in many ways as a spring does not have any mass and its cross section is not explicitly included. A more complicated but realistic element is the bar element as shown in Figure 6.4, which is a uniform rod with a cross section area A, Young's elastic modulus E, and a length L. A bar element can only support tension and compression, it cannot support bending. For this reason, it is also called a truss element.

The displacements at nodes i and j are u_i and u_j, respectively. The forces at the corresponding nodes are f_i and f_j. Now we have to derive its stiffness matrix. Assuming the bar is linearly elastic, the stress σ is thus related to strain ϵ via $\sigma = E\epsilon$. Since $\epsilon = (u_j - u_i)/L$ and $\sigma = f/A$ where F is the force in the bar element, we have

$$f = \frac{EA}{L}(\Delta u) = k(\Delta u), \tag{6.17}$$

where $\Delta u = u_j - u_i$ is the extension or elongation of the bar element. Now the equivalent spring stiffness constant is

$$k = \frac{EA}{L}. \tag{6.18}$$

Therefore, the stiffness matrix $\mathbf{K}$ for this bar becomes

$$\mathbf{K} = \begin{pmatrix} k & -k \\ -k & k \end{pmatrix} = \frac{EA}{L} \begin{pmatrix} 1 & -1 \\ -1 & 1 \end{pmatrix}. \tag{6.19}$$

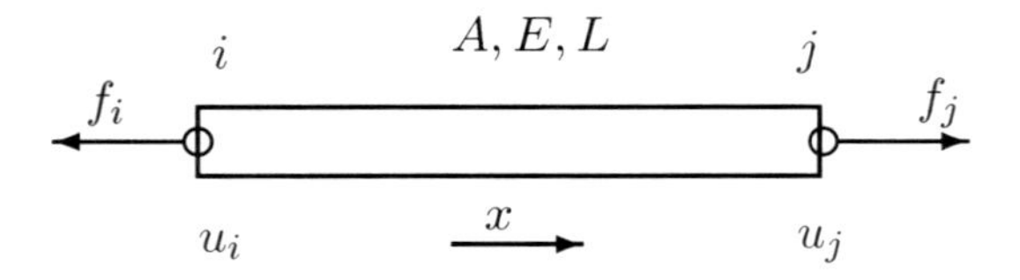

Figure 6.4: Bar element.

We have up to now only discussed 1-D systems where all displacements u_i or u_j are along the bar direction, and each node has only one displacement (one degree of freedom). We now extend to study 2-D systems. In 2-D, each node i has two displacements u_i (along the bar direction) and v_i (perpendicular to the bar direction). Thus, each node has two degrees of freedom.

If we rotate the bar element by an angle θ as shown in Figure 6.5, we cannot use the standard addition to assemble the system. A transformation is needed between the global coordinates (x, y) to the local coordinates (x', y'). From the geometrical consideration, the global displacements u_i and v_i at node i are related to the local displacement u_i' and (usually) $v_i' = 0$.

$$\begin{pmatrix} u_i' \\ v_i' \end{pmatrix} = \begin{pmatrix} \cos\theta & \sin\theta \\ -\sin\theta & \cos\theta \end{pmatrix} \begin{pmatrix} u_i \\ v_i \end{pmatrix}. \tag{6.20}$$

Using the similar transformation for u_j and v_j, we get the transformation for the two-node bar element

$$\mathbf{u}' = \begin{pmatrix} u_i' \\ v_i' \\ u_j' \\ v_j' \end{pmatrix} = \begin{pmatrix} \cos\theta & \sin\theta & 0 & 0 \\ -\sin\theta & \cos\theta & 0 & 0 \\ 0 & 0 & \cos\theta & \sin\theta \\ 0 & 0 & -\sin\theta & \cos\theta \end{pmatrix} \begin{pmatrix} u_i \\ v_i \\ u_j \\ v_j \end{pmatrix},$$

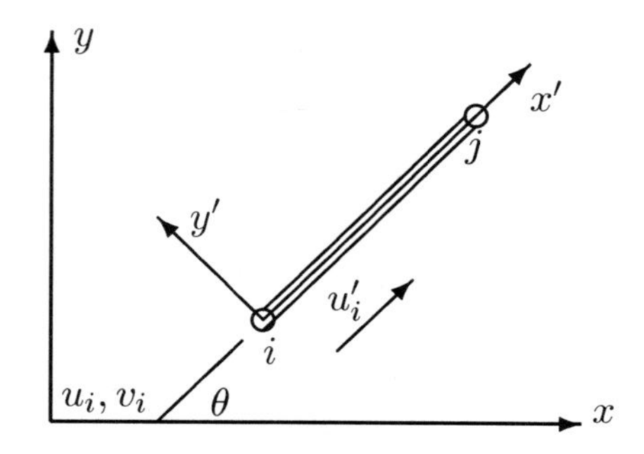

Figure 6.5: 2-D transformation of coordinates.

which can be written as

$$\mathbf{u}' = \mathbf{R}\mathbf{u}, \tag{6.21}$$

where

$$\mathbf{R} == \begin{pmatrix} \cos\theta & \sin\theta & 0 & 0 \\ -\sin\theta & \cos\theta & 0 & 0 \\ 0 & 0 & \cos\theta & \sin\theta \\ 0 & 0 & -\sin\theta & \cos\theta \end{pmatrix}. \tag{6.22}$$

The same applies to transform the force,

$$\mathbf{f}' = \mathbf{R}\mathbf{f}, \tag{6.23}$$

and the stiffness matrix in local coordinates is

$$\mathbf{K}'\mathbf{u}' = \mathbf{f}'. \tag{6.24}$$

As the calculation is mainly based on the global coordinates, and the assembly should be done by transforming the local systems to the global coordinates, by combining the above two equations, we have

$$\mathbf{K}'\mathbf{R}\mathbf{u} = \mathbf{R}\mathbf{f}, \tag{6.25}$$

or

$$\mathbf{R}^{-1}\mathbf{K}'\mathbf{R}\mathbf{u} = \mathbf{K}\mathbf{u} = \mathbf{f}, \tag{6.26}$$

which is equivalent to a global stiffness matrix

$$\mathbf{K} = \mathbf{R}^{-1}\mathbf{K}'\mathbf{R}. \tag{6.27}$$

The stiffness matrix $\mathbf{K}$ is a 4×4 matrix in 2-D.

Bar elements can only elongate or shrink, they do not support bending or deflection. For bending, we need the beam elements which include a rotation around the end nodes θ_i and θ_j. In this case, each node has three degrees of freedom (u_i, v_i, θ_i), and the stiffness matrix is therefore a 6×6 matrix in 2-D. The concept of a beam element is based on the Euler's beam bending equation

$$M = EI\frac{\partial^2 v}{\partial x^2}, \tag{6.28}$$

where M is the bending moment, E is the Young's modulus, and I is the second moment of area. Thus, the stiffness matrix of a 2-D beam element is

$$\mathbf{K} = \begin{pmatrix} \alpha & 0 & 0 & -\alpha & 0 & 0 \\ 0 & \beta & \gamma & 0 & -\beta & \gamma \\ 0 & \beta L/2 & 2\gamma L/3 & 0 & -\beta L/2 & \gamma L/3 \\ -\alpha & 0 & 0 & \alpha & 0 & 0 \\ 0 & -\beta & -\gamma & 0 & \beta & -\gamma \\ 0 & \beta L/2 & \gamma L/3 & 0 & -\beta L/2 & 2\gamma L/3 \end{pmatrix}, \tag{6.29}$$

where

$$\alpha \equiv \frac{AE}{L}, \qquad \beta \equiv \frac{12EI}{L^3}, \qquad \gamma \equiv \frac{6EI}{L^2}. \tag{6.30}$$

The stiffness matrix acts on the following degrees of freedom $(u_i, v_i, \theta_i, u_j, v_j, \theta_j)$.

For more complicated elements, it is more convenient and even necessary to use a formal approach in terms of shape functions and weak formulations. Figure 6.6 shows several common elements.

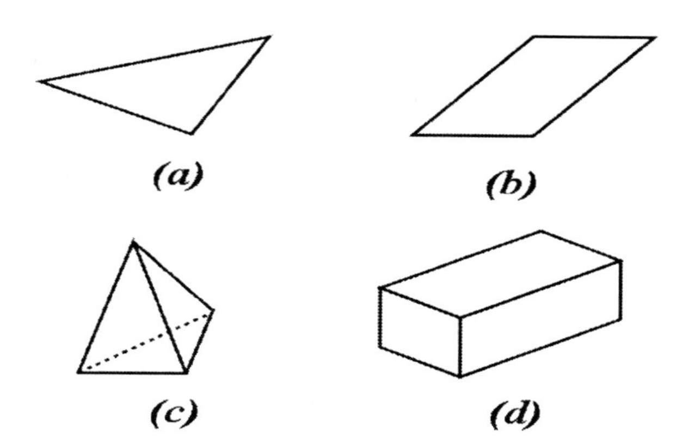

Figure 6.6: Common elements: (a) triangular element; (b) quadrilateral element; (c) tetrahedron; and (d) hexahedron.

6.2 Finite Element Formulation

6.2.1 Weak Formulation

Many problems are modelled in terms of partial differential equations, which can generally be written as

$$\mathcal{L}(u) = 0, \qquad \mathbf{x} \in \Omega, \tag{6.31}$$

where $\mathcal{L}$ is a differential operator, often linear. This problem is usually completed with the essential boundary condition (or prescribed values $\bar{u}$), $\mathcal{E}(u) = (u - \bar{u}) = 0$ for $\mathbf{x} \in \partial\Omega_E$, and natural boundary conditions $\mathcal{B}(u) = 0$ for $\mathbf{x} \in \partial\Omega_N$. Natural boundary conditions are usually concerned with flux or force.

Assuming that the true solution u can be approximated by u_h over a finite element mesh with an averaged element size or mean distance h between two adjacent nodes, the above equation can be approximated as

$$\mathcal{L}(u_h) \approx 0. \tag{6.32}$$

The ultimate goal is to construct a method of computing u_h such that the error $|u_h - u|$ is minimized. Generally speaking, the residual $R(u_1, ..., u_M, \mathbf{x}) = \mathcal{L}(u_h(\mathbf{x}))$ varies with space and time. There are several methods to minimize R. Depending on the scheme of minimization and the choice of shape functions, various methods can be formulated. These include the weighted residual method, the method of least squares, the Galerkin method and others.

Multiplying both sides of equation (6.32) by a test function or a proper weighting function w_i, integrating over the domain and using associated boundary conditions, we can write the general weak formulation of Zienkiewicz-type as

$$\int_\Omega \mathcal{L}(u_h) w_i d\Omega$$

$$+ \int_{\partial\Omega_N} \mathcal{B}(u_h)\bar{w}_i d\Gamma + \int_{\partial\Omega_E} \mathcal{E}(u_h)\tilde{w}_i d\Gamma_E \approx 0, \tag{6.33}$$

where $(i = 1, 2, ..., M)$, and $\bar{w}_i$ and $\tilde{w}_i$ are the values of w_i on the natural and essential boundaries. If we can approximate the solution u_h by the expansion in term of shape function N_i

$$u_h(u, t) = \sum_{i=1}^{M} u_i(t) N_i(x) = \sum_{j=1}^{M} u_j N_j, \tag{6.34}$$

it requires that $N_i = 0$ on $\partial\Omega_E$ so that we can choose $\tilde{w}_i = 0$ on $\partial\Omega_E$. Thus, only the natural boundary conditions are included since the essential boundary conditions are automatically satisfied. In addition, there is no much limitation on the choice of w_i and $\bar{w}_i$. If we choose $\bar{w}_i = -w_i$ on the natural boundary so as to simplify the formulation, we have

$$\int_\Omega \mathcal{L}(u_h) w_i d\Omega \approx \int_{\partial\Omega_N} \mathcal{B}(u_h) w_i d\Gamma. \tag{6.35}$$

6.2.2 Galerkin Method

There are many different ways to choose the test functions w_i and shape functions N_i. One of the most popular methods is the Galerkin method where the test functions are the same as the shape functions, or $w_i = N_i$. In this special case, the formulation simply becomes

$$\int_\Omega \mathcal{L}(u_h) N_i d\Omega \approx \int_{\partial\Omega_N} \mathcal{B}(u_h) N_i d\Gamma. \tag{6.36}$$

The discretization of this equation will usually lead to an algebraic matrix equation.

On the other hand, if we use the Dirac delta function as the test functions $w_i = \delta(\mathbf{x} - \mathbf{x}_i)$, the method is called the collocation method which uses the interesting properties of the Dirac function

$$\int_\Omega f(\mathbf{x})\delta(\mathbf{x} - \mathbf{x}_i) d\Omega = f(\mathbf{x}_i), \tag{6.37}$$

together with $\delta(\mathbf{x} - \mathbf{x}_i) = 1$ at $\mathbf{x} = \mathbf{x}_i$ and $\delta(\mathbf{x} - \mathbf{x}_i) = 0$ at $\mathbf{x} \neq \mathbf{x}_i$.

6.2.3 Shape Functions

The main aim of the finite element method is to find an approximate solution $u_h(\mathbf{x}, t)$ for the exact solution u on some nodal points,

$$u_h(\mathbf{x}, t) = \sum_{i=1}^{M} u_i(t) N_i(\mathbf{x}) \tag{6.38}$$

where u_i are unknown coefficients or the values of u at the discrete nodal point i. Functions N_i $(i = 1, 2, ..., M)$ are linearly independent functions that vanish on the part of the essential boundary. At any node i, we have $N_i = 1$, and $N_i = 0$ at any other nodes, or

$$\sum_{i=1}^{M} N_i = 1, \qquad N_i(\mathbf{x}_j) = \delta_{ij}. \tag{6.39}$$

The functions $N_i(\mathbf{x})$ are referred to as basis functions, trial functions or more often shape functions in the literature of finite element methods.

Linear Shape Functions

For the simplest 1-D element with two nodes i and j, the linear shape functions (shown in Figure 6.7.) can be written as

$$N_i = 1 - \frac{x}{L} = \frac{1-\xi}{2}, N_j = \xi = \frac{x}{L} = \frac{1+\xi}{2}, \tag{6.40}$$

where ξ is the natural coordinate

$$\xi = \frac{x - x_o}{L/2}, \qquad L = |x_j - x_i|, \qquad x_o = \frac{x_i + x_j}{2}, \tag{6.41}$$

where x_o is the midpoint of the element, and $\xi_i = -1$ at $x = x_i$ and $\xi_j = 1$ at $x = x_j$.

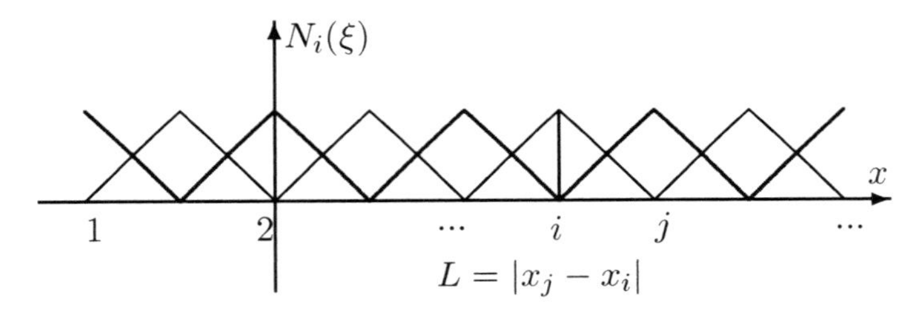

Figure 6.7: The 1-D linear shape functions.

A linear shape function spans only two adjacent nodes i and j, and its requires two coefficients in the generic form

$$N(\xi) = a + b\xi. \tag{6.42}$$

Quadratic Shape Functions

Suppose we want to get higher order approximations, we can use, say, the quadratic shape functions which span three adjacent nodes i, j, and k. Three coefficients needs to be determined

$$N(\xi) = a + b\xi + c\xi^2. \tag{6.43}$$

Using the conditions $\xi_i = -1$ at $x = x_i$ and $\xi_j = 1$ at $x = x_j$, and the known displacements u_i, u_j and u_k, we have

$$u_i = a + b(-1) + c(-1)^2, \tag{6.44}$$

$$u_j = a, \tag{6.45}$$

and

$$u_k = a + b(1) + c(1)^2, \tag{6.46}$$

whose solutions are

$$\begin{pmatrix} a \\ b \\ c \end{pmatrix} = \begin{pmatrix} u_j \\ \frac{1}{2}(u_i - 2u_j + u_k) \\ \frac{1}{2}(u_k - u_i) \end{pmatrix}. \tag{6.47}$$

Substituting this into equation (6.43), we have

$$u = \frac{\xi(\xi-1)}{2} u_i + (1-\xi^2) u_j + \frac{\xi(\xi+1)}{2} u_k, \tag{6.48}$$

which is equivalent to

$$u = N_i u_i + N_j u_j + N_k u_k, \tag{6.49}$$

where

$$\mathbf{N} = [N_i, N_j, N_k] = [\frac{\xi(\xi-1)}{2}, (1-\xi^2), \frac{\xi(\xi+1)}{2}]. \tag{6.50}$$

Lagrange Polynomials

The essence of the shape functions is the interpolation, and the interpolation functions can be many different types. Lagrange polynomials are popularly used to construct shape functions. The $n-1$ order Lagrange polynomials require n nodes, and the associated shape functions can generally be written as

$$N_i(\xi) = \Pi_{j=1, j\neq i}^{n} \frac{(\xi - \xi_j)}{(\xi_i - \xi_j)}$$

$$= \frac{(\xi-\xi_1)...(\xi-\xi_{i-1})(\xi-\xi_{i+1})...(\xi-\xi_n)}{(\xi_i-\xi_1)...(\xi_i-\xi_{i-1})(\xi_i-\xi_{i+1})...(\xi_i-\xi_n)}, \tag{6.51}$$

where ξ_j means that value of ξ at node j. For $n = 3$, it is straightforward to validate that

$$N_1(\xi) = \frac{\xi(\xi-1)}{2}, \quad N_2(\xi) = 1-\xi^2, \quad N_3(\xi) = \frac{\xi(\xi+1)}{2}. \tag{6.52}$$

This method of formulating shape functions can be easily extended to 2D and 3D cases and for isoparametric elements. The derivative of $N_i(x)$ with respect to ξ is given by

$$N_i'(\xi) = \sum_{k=1,k\neq i}^{n} \frac{1}{(\xi_i - \xi_j)} \Pi_{j=1,j\neq i}^{n} \frac{(\xi - \xi_j)}{(\xi_i - \xi_j)}. \tag{6.53}$$

2D Shape Functions

The shape functions we discussed earlier are 1D shape functions (with one independent variables x or ξ) for 1D elements. For 2D elements such as quadrilateral elements, corresponding shape functions with two independent variables: x and y, or ξ and η. Using the natural coordinates ξ and η shown in Fig. 6.8, we can construct various shape functions.

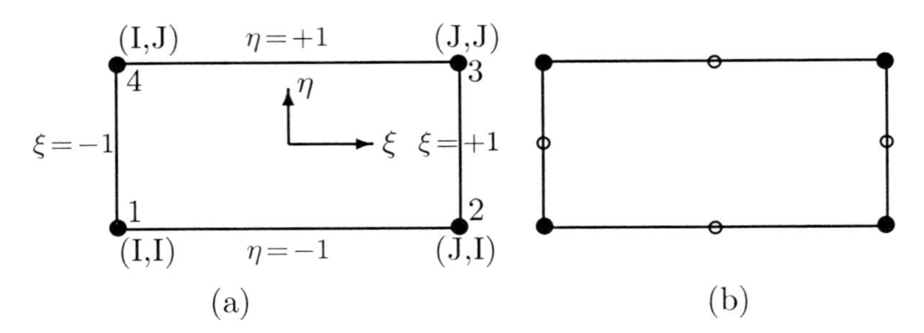

Figure 6.8: (a) A bilinear quadrilateral element; (b) A quadratic quadrilateral element.

For a bilinear quadrilateral (Q4) element, we use bilinear approximations for the displacement field u and v. If we use

$$u = \alpha_0 + \alpha_1 x + \alpha_2 y + \alpha_3 xy, \tag{6.54}$$

$$v = \beta_0 + \beta_1 x + \beta_2 y + \beta_3 xy, \tag{6.55}$$

and express them in terms of shape function N_i

$$u = \sum N_i u_i, \qquad v = \sum N_j v_j, \tag{6.56}$$

we can derive the shape functions by following the similar procedure as discussed above. We have

$$N_1 = \frac{(1-\xi)(1-\eta)}{4}, \qquad N_2 = \frac{(1+\xi)(1-\eta)}{4}, \tag{6.57}$$

$$N_3 = \frac{(1+\xi)(1+\eta)}{4}, \qquad N_4 = \frac{(1-\xi)(1+\eta)}{4}, \tag{6.58}$$

From the 1-D linear shape functions

$$N_I^{(2)}(\xi) = \frac{(1-\xi)}{2}, \qquad N_J^{(2)}(\xi) = \frac{(1+\xi)}{2}, \tag{6.59}$$

for a 2-node element (along x) where the superscript '(2)' means 2 nodes, we can also write another set of linear shape functions for a 2-node element in the y-direction. We have

$$N_I^{(2)}(\eta) = \frac{(1-\eta)}{2}, \qquad N_J^{(2)}(\eta) = \frac{(1+\eta)}{2}. \tag{6.60}$$

If we label the nodes by a pair (I, J) in 2D coordinates, we have

$$N_i(\xi, \eta) = N_{IJ} = N_I^{(2)} N_J^{(2)}. \tag{6.61}$$

We can see that

$$N_1(\xi,\eta) = N_I^{(2)}(\xi) N_I^{(2)}(\eta), \; N_2(\xi,\eta) = N_J^{(2)}(\xi) N_I^{(2)}(\eta), \tag{6.62}$$

and

$$N_3(\xi,\eta) = N_J^{(2)}(\xi) N_J^{(2)}(\eta), \; N_4(\xi,\eta) = N_I^{(2)}(\xi) N_J^{(2)}(\eta). \tag{6.63}$$

In fact, higher-order shape functions for 2D and 3D elements can be systematically derived this way.

If we approximate the displacement field using higher-order approximations, then we are dealing with the quadratic quadrilateral (Q8) element because we have to use eight nodes (4 finite element nodes and 4 midpoints). In this case, the shape functions are much more complicated, for example, the shape function N_2 becomes

$$N_2 = \frac{(1-\xi)(1-\eta)}{4} - \frac{1}{4}[(1-\xi^2)(1-\eta) + (1+\xi)(1-\eta^2)]. \tag{6.64}$$

Derivatives

Using the assumptions that $u_i(t)$ does not depend on space and $N_i(\mathbf{x})$ does not depend on time, the derivatives of u can be approximated as

$$\frac{\partial u}{\partial \mathbf{x}} \approx \frac{\partial u_h}{\partial \mathbf{x}} = \sum_{i=1}^{M} u_i(t) N'(\mathbf{x}),$$

$$\dot{u} \approx \frac{\partial u_h}{\partial t} = \sum_{i=1}^{M} \dot{u}_i N(\mathbf{x}), \tag{6.65}$$

where we have used the notations: $' = d/d\mathbf{x}$ and $\dot{} = \frac{\partial}{\partial t}$. The derivatives of the shape functions N_i can be calculated in the similar manner as shown in 6.53. Higher order derivatives are then calculated in a similar way.

Gauss Quadrature

In the finite element analysis, the calculation of stiffness matrix and application of boundary conditions such as in equation (6.36) involve the integration over elements. Such numerical integration is often carried out in terms of natural coordinates ξ and η, and the Gauss integration or Gauss quadrature as discussed in Section 1.1.6 is usually used for evaluating integrals numerically. Gauss quadrature has relatively high accuracy. For example, the n-point Gauss quadrature for one-dimensional integrals

$$\mathcal{I} = \int_{-1}^{1} \psi(\xi) d\xi \approx \sum_{i=1}^{n} w_i \psi_i. \tag{6.66}$$

For the case of $n = 3$, we have

$$\int_{-1}^{1} \psi(\xi) d\xi \approx \sum_{i=1}^{3} w_i \psi_i = \frac{1}{9}[8\psi_2 + 5(\psi_1 + \psi_3)], \tag{6.67}$$

which is schematically shown in Figure 6.9.

For two-dimensional integrals, we use n^2-point Gauss quadrature of order n, and we have

$$\mathcal{I} = \int_{-1}^{1}\int_{-1}^{1} \psi(\xi,\eta)d\xi d\eta = \sum_{i=1}^{n}\sum_{j=1}^{n} w_i w_j \psi_{ij}, \tag{6.68}$$

where $\psi_{ij} = \psi(\xi_i, \eta_j)$. In the case of $n = 3$, we have 9 points (shown in Figure 6.9), and the quadrature becomes

$$\mathcal{I} = \int_{-1}^{1}\int_{-1}^{1} \psi(\xi,\eta) \approx \sum_{i=1}^{3}\sum_{j=1}^{3} w_i w_j \psi_{i+3*(j-1)}(\xi_i,\eta_j)$$

$$= \frac{1}{81}[25(\psi_1+\psi_3+\psi_7+\psi_9)+64\psi_5+40(\psi_2+\psi_4+\psi_6+\psi_8)]. \tag{6.69}$$

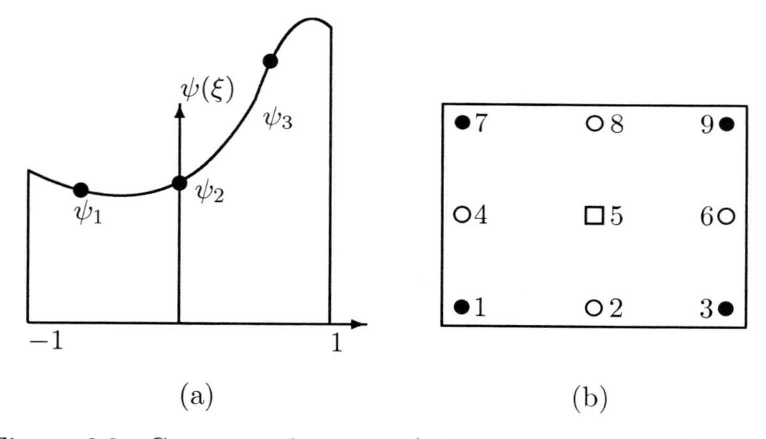

Figure 6.9: Gauss quadrature: a) 1-D integration with $|\xi_1 - \xi_2| = |\xi_2 - \xi_3| = \sqrt{3/5}$; and b) 2-D 9-point integration over a quadrilateral element with point 3 at $(\xi_3, \eta_1) = (\sqrt{3/5}, -\sqrt{3/5})$ and point 9 at (ξ_3, η_3)=$(\sqrt{3/5}, \sqrt{3/5})$.

6.2.4 Variational Formulations

Finite element analysis has the rigorous mathematical background based on the calculus of variations. This connection can easily be demonstrated using an example. From the calculus of variations, we can see that the stiffness equation

$$\mathbf{K}\mathbf{u} = \mathbf{b}, \tag{6.70}$$

is in fact equivalent to make stationary the following quadratic functional Π

$$\Pi(\mathbf{u}) = \frac{1}{2}\mathbf{u}^T\mathbf{K}\mathbf{u} - \mathbf{u}^T\mathbf{b}. \tag{6.71}$$

The stationary condition requires that

$$\frac{\delta\Pi}{\delta\mathbf{u}} = 0. \tag{6.72}$$

From

$$\delta\Pi = \frac{1}{2}[\delta\mathbf{u}^T\mathbf{K}\mathbf{u} + \mathbf{u}^T\mathbf{K}\delta\mathbf{u}] - \delta\mathbf{u}^T\mathbf{b}, \tag{6.73}$$

and the symmetry of $\mathbf{K}$ (that is $\delta\mathbf{u}^T\mathbf{K}\mathbf{u} = \mathbf{u}^T\mathbf{K}\delta\mathbf{u}^T$), we have

$$\delta\Pi = \delta\mathbf{u}^T(\mathbf{K}\mathbf{u} - \mathbf{b}), \tag{6.74}$$

which leads to

$$\frac{\delta\Pi}{\delta\mathbf{u}} = \mathbf{K}\mathbf{u} - \mathbf{b} = 0. \tag{6.75}$$

If we start from the differential equation, we will get the same results. As another example, now let us solve the 1-D Poisson equation

$$\alpha\frac{d^2u(x)}{dx^2} + f(x) = 0, \qquad x \in [a, b], \tag{6.76}$$

with homogeneous boundary conditions $u(a) = u(b) = 0$. This is a boundary value problem. The finite element method can be viewed as a variational method if we consider it as an optimisation problem.

Let $u^*(x)$ be the optimal or true solution that satisfies all the appropriate boundary conditions, and let $u(x)$ is an estimate to $u^*(x)$. If we define the linear weak form (u, v)

$$(u, v) = \int_a^b u(x)v(x)dx, \tag{6.77}$$

and the functional $J(u)$

$$\Pi(u) = \frac{\alpha}{2}(u', u') - (f, u)$$

$$= \frac{1}{2}\int_a^b \frac{du}{dx}\alpha\frac{du}{dx}dx - \int_a^b f(x)u(x)dx, \tag{6.78}$$

then $(u', v') = (f, v)$ is equivalent to virtual work where v is continuous on $[a, b]$. The minimisation $\Pi(u^*) \leq \Pi(u)$ corresponds to the minimisation of the total potential energy (or more precisely, the action) because the first term of J is the kinetic energy T and the second term of the potential energy V.

$$T = \frac{1}{2}\int_a^b \alpha(u')^2 dx, \qquad V = -\int_a^b f(x)u(x)dx. \tag{6.79}$$

Variational Forms

Let us define the inner product of two functions

$$(u, v) = \int_\Omega uv d\Omega. \tag{6.80}$$

If one function is given, say, $u = f(\boldsymbol{x})$ is known, then

$$L(v) = (f, v) = \int_\Omega v f(\boldsymbol{x}) d\Omega. \tag{6.81}$$

The form $L(v)$ is said to be linear if it satisfies $L(\alpha v + \beta w) = \alpha L(v) + \beta L(w)$ for any two arbitrary coefficients α and β. All the linear forms span a linear space $\mathcal{H}$. If a linear space $\mathcal{H}$ together with an inner product defined on it is called an inner-product space with a norm $\|v\|_{\mathcal{H}} = \sqrt{(v, v)}$. If Ω is a bounded smooth domain in $\Re^n$ where n is the number of dimensions, then an L^2 space, or $L^2(\Omega)$, is the set of square-integrable functions on Ω,

$$L^2(\Omega) = \{v \Big| \int_\Omega v^2 d\Omega < \infty\}. \tag{6.82}$$

A Hilbert space is the space $\mathcal{H} = L^2(\Omega)$ equipped with the inner product $(u, v)_{L^2} = \int_\Omega uv d\Omega$ and the associated norm

$$\|v\|_{L^2} = \sqrt{(v, v)_{L^2}}. \tag{6.83}$$

Variational Forms of FE

Now let us study the linear differential equation

$$\mathcal{L}(u) = 0. \tag{6.84}$$

Writing it in the integral form and using the expansion $u_h = \sum_{j=1}^{M} u_j N_j$, we have

$$\int_\Omega \mathcal{L}(u_h) N_i d\Omega = 0, \tag{6.85}$$

where N_i functions then span a vector space $\mathcal{V}$ with a finite dimension $M = \dim(\mathcal{V})$.

The Galerkin method is equivalent to find $u_h \in \mathcal{V}$ so that

$$(\mathcal{L}(u_h), v) = 0, \qquad \forall v \in \mathcal{V}. \tag{6.86}$$

If $\mathcal{L}$ is a second-order (or higher) differential operator, then its integral form with all test function v and integration by parts will leads to the generic form

$$a(u_h, v) = L(v), \tag{6.87}$$

where $a(u_h, v)$ is a bilinear operator obtained by integration by parts. For example, for $\mathcal{L}(u_h) = u_h'' - f(x) = 0$ on $[a, b]$, we have

$$a(u_h, v) = \int_a^b u_h' v' dx = L(v) = -\int_a^b f v dx. \tag{6.88}$$

Particularly, we have the limit $u_h \to u$ when $M \to \infty$. Then, we have

$$a(u, v) = L(v), \qquad \forall u, v \in \mathcal{V}. \tag{6.89}$$

Such a variational formulation is also referred to as the weak formulation or simply the weak form. Loosely speaking, the finite element method formulated this way has the following a priori error estimate:

$$\|u - u_h\| = \|\nabla u - \nabla u_h\| \le \Lambda h^s |u|_{L^2}, \tag{6.90}$$

where $\Lambda > 0$ is a constant. s is the order of the basis functions and h is the element size.

For the Poisson equation

$$\nabla \cdot (k \nabla u) + f(\boldsymbol{x}) = 0, \qquad \boldsymbol{x} \in \Omega, \tag{6.91}$$

with a Robin type of natural boundary conditions

$$u = 0, \qquad \boldsymbol{x} \in \partial \Omega_E,$$

$$k \frac{\partial u}{\partial n} + q(u - \bar{u}) = 0, \qquad \boldsymbol{x} \in \partial \Omega_N, \tag{6.92}$$

we can integrate it by parts, and we then have

$$a(u, v) = \int_\Omega k \nabla u \nabla v d\Omega + \int_{\partial \Omega_N} quv d\Gamma, \tag{6.93}$$

and

$$L(v) = \int_\Omega f(\boldsymbol{x}) v d\Omega + \int_{\partial \Omega_N} q\bar{u} v d\Gamma. \tag{6.94}$$

For a transient problem,

$$\frac{\partial^2 u}{\partial t^2} = \nabla \cdot (k \nabla u), \tag{6.95}$$

the weak form becomes

$$(\frac{\partial^2 u}{\partial t^2}, v) + a(u, v) = L(v), \tag{6.96}$$

where the extra time-dependent term should be discretized using time-stepping schemes, which will be discussed in detail in later chapters.

Chapter 7

Elasticity

7.1 Hooke's Law and Elasticity

The basic Hooke's law of elasticity concerns an elastic body such as a spring, and it states that the extension x is proportional to the load F, that is

$$F = kx, \tag{7.1}$$

where k the spring constant. However, this equation only works for 1-D deformations. For a bar of uniform cross-section with a length L and a cross section area A, it is more convenient to use strain ε and stress σ. The stress and strain are defined by

$$\sigma = \frac{F}{A}, \qquad \varepsilon = \frac{\Delta L}{L}, \tag{7.2}$$

where ΔL is the extension. The unit of stress is $\mathrm{N/m^2}$, while the strain is dimensionless, though it is conventionally expressed in m/m or % (percentage) in engineering. For the elastic bar, the stress-strain relationship is

$$\sigma = E\varepsilon, \tag{7.3}$$

where E is the Young's modulus of elasticity. Written in terms F and $x = \Delta L$, we have

$$F = \frac{EA}{L}\Delta L = kx, \qquad k = \frac{EA}{L}, \tag{7.4}$$

where k is the equivalent spring constant for the bar. This equation is still only valid for any unidirectional compression or extension. For the 2-D and 3-D deformation, we need to generalize Hooke's law. For the general stress tensor (also called Cauchy stress tensor)

$$\boldsymbol{\sigma} = \begin{pmatrix} \sigma_{xx} & \sigma_{xy} & \sigma_{xz} \\ \sigma_{yx} & \sigma_{yy} & \sigma_{yz} \\ \sigma_{zx} & \sigma_{zy} & \sigma_{zz} \end{pmatrix} = \begin{pmatrix} \sigma_{11} & \sigma_{12} & \sigma_{13} \\ \sigma_{21} & \sigma_{22} & \sigma_{23} \\ \sigma_{31} & \sigma_{32} & \sigma_{33} \end{pmatrix}, \tag{7.5}$$

and strain tensor

$$\boldsymbol{\varepsilon} = \begin{pmatrix} \varepsilon_{xx} & \varepsilon_{xy} & \varepsilon_{xz} \\ \varepsilon_{yx} & \varepsilon_{yy} & \varepsilon_{yz} \\ \varepsilon_{zx} & \varepsilon_{zy} & \varepsilon_{zz} \end{pmatrix} = \begin{pmatrix} \varepsilon_{11} & \varepsilon_{12} & \varepsilon_{13} \\ \varepsilon_{21} & \varepsilon_{22} & \varepsilon_{23} \\ \varepsilon_{31} & \varepsilon_{32} & \varepsilon_{33} \end{pmatrix}, \tag{7.6}$$

it can be proved later that these tensors are symmetric, that is $\boldsymbol{\sigma} = \boldsymbol{\sigma}^T$ and $\boldsymbol{\varepsilon} = \boldsymbol{\varepsilon}^T$, which leads to

$$\sigma_{xy} = \sigma_{yx}, \quad \sigma_{xz} = \sigma_{zx}, \quad \sigma_{yz} = \sigma_{zy}, \tag{7.7}$$

and

$$\varepsilon_{xy} = \varepsilon_{yx}, \quad \varepsilon_{xz} = \varepsilon_{zx}, \quad \varepsilon_{yz} = \varepsilon_{zy}. \tag{7.8}$$

Therefore, we only have 6 independent components or unknowns for stresses and 6 unknown strain components.

The strain tensor is defined by the displacement $\mathbf{u}^T = (u_1, u_2, u_3)$

$$\varepsilon_{ij} = \frac{1}{2}\left(\frac{\partial u_i}{\partial x_j} + \frac{\partial u_j}{\partial x_i}\right), \tag{7.9}$$

where $x_1 = x$, $x_2 = y$, and $x_3 = z$. Sometimes, it is useful to write

$$\boldsymbol{\varepsilon} = \frac{1}{2}(\nabla \boldsymbol{u} + \nabla \boldsymbol{u}^T). \tag{7.10}$$

The generalized Hooke's law can be written as

$$\varepsilon_{xx} = \frac{1}{E}[\sigma_{xx} - \nu(\sigma_{yy} + \sigma_{zz})], \tag{7.11}$$

$$\varepsilon_{yy} = \frac{1}{E}[\sigma_{yy} - \nu(\sigma_{xx} + \sigma_{zz})], \tag{7.12}$$

$$\varepsilon_{zz} = \frac{1}{E}[\sigma_{zz} - \nu(\sigma_{xx} + \sigma_{yy})], \tag{7.13}$$

$$\varepsilon_{xy} = \frac{1+\nu}{E}\sigma_{xy}, \tag{7.14}$$

$$\varepsilon_{xz} = \frac{1+\nu}{E}\sigma_{xz}, \tag{7.15}$$

$$\varepsilon_{yz} = \frac{1+\nu}{E}\sigma_{yz}, \tag{7.16}$$

where ν is the Poisson's ratio, and it measures the tendency of extension in transverse directions (say, x and y) when the elastic body is stretched in one direction (say, z). It can be defined as the ratio of the transverse contract strain (normal to the applied load) to the axial strain in a stretched cylindrical bar in the direction of the applied force. For a perfectly incompressible material, $\nu = 0.5$, and $\nu = 0 \sim 0.5$ for most common materials. For example, steels have $\nu = 0.25 \sim 0.3$. Some auxetic material such as polymer foams or anti-rubbers have a negative Poisson's ratio $\nu < 0$.

This generalized Hooke's law can concisely be written as

$$\varepsilon_{ij} = \frac{1+\nu}{E}\sigma_{ij} - \frac{\nu}{E}\sigma_{kk}\delta_{ij}, \tag{7.17}$$

where we have used the Einstein's summation convention $\sigma_{kk} = \sigma_{xx} + \sigma_{yy} + \sigma_{zz}$. Another related quantity is the pressure, which is defined by

$$p = -\frac{1}{3}\sigma_{kk} = -\frac{\sigma_{xx} + \sigma_{yy} + \sigma_{zz}}{3}. \tag{7.18}$$

The negative sign comes from the conventions that a positive normal stress results in tension, and negative one in compression, while the positive pressure acts in compression. Sometimes, it is more convenient to express the stress tensor in terms of pressure and devitoric stress tensor s_{ij}

$$\sigma_{ij} = -p\delta_{ij} + s_{ij}. \tag{7.19}$$

If we want to invert equation (7.17), we have first express σ_{kk} in terms of ε_{kk} so that the right hand side of the new expression does not contain the stress σ_{kk}. By contraction using $j \to i$, we have

$$\varepsilon_{ii} = \frac{1+\nu}{E}\sigma_{ii} - \frac{\nu}{E}\sigma_{kk}\delta_{ii} = \frac{1-2\nu}{E}\sigma_{ii}, \tag{7.20}$$

where we have used $\delta_{ii} = \delta_{11} + \delta_{22} + \delta_{33} = 1 + 1 + 1 = 3$ and $\sigma_{ii} = \sigma_{kk}$. In engineering, the quantity

$$\varepsilon_{kk} = \varepsilon_{xx} + \varepsilon_{yy} + \varepsilon_{zz} = \frac{\partial^2 u_1}{\partial x^2} + \frac{\partial^2 u_2}{\partial y^2} + \frac{\partial^2 u_3}{\partial z^2} = \nabla \cdot \boldsymbol{u}, \tag{7.21}$$

means the fractional change in volume, known as the dilation. This gives that

$$\sigma_{ii} = \sigma_{kk} = \frac{E}{1-2\nu}\varepsilon_{kk}. \tag{7.22}$$

Substituting it into equation (7.17), we have

$$\varepsilon_{ij} = \frac{1+\nu}{E}\sigma_{ij} - \frac{\nu}{E}(\frac{E}{1-2\nu}\varepsilon_{kk})\delta_{ij}, \tag{7.23}$$

or after some rearrangement

$$\frac{1+\nu}{E}\sigma_{ij} = \varepsilon_{ij} + \frac{\nu}{1-2\nu}\varepsilon_{kk}\delta_{ij}, \tag{7.24}$$

which can be written as

$$\sigma_{ij} = 2G\varepsilon_{ij} + \lambda\varepsilon_{kk}\delta_{ij}, \tag{7.25}$$

where μ and λ are Lamé constants. They are

$$G = \mu = \frac{E}{2(1+\nu)}, \qquad \lambda = \frac{\nu E}{(1+\nu)(1-2\nu)}. \tag{7.26}$$

This stress-strain relationship can also be written as

$$\boldsymbol{\sigma} = 2G\boldsymbol{\varepsilon} + \lambda(\nabla \cdot \boldsymbol{u})\boldsymbol{\delta}. \tag{7.27}$$

In engineering, $G = \mu$ is called the shear modulus, while $K = \frac{E}{3(1-2\nu)}$ is called the bulk modulus which is the ratio of pressure $-p$ to the volume change rate $\Delta V/V$.

7.2 Plane Stress and Plane Strain

In engineering applications and finite element analysis, the stress tensor $\boldsymbol{\sigma}$ and strain tensor $\boldsymbol{\epsilon}$ are not written as tensor forms but vector forms $\boldsymbol{\sigma} = (\sigma_{xx}, \sigma_{yy}, \sigma_{zz}, \sigma_{xy}, \sigma_{yz}, \sigma_{zx})^T$, and $\boldsymbol{\epsilon} = (\epsilon_{xx}, \epsilon_{yy}, \epsilon_{zz}, \gamma_{xy}, \gamma_{yz}, \gamma_{zx})^T$. The strain tensor is usually defined as

$$\varepsilon_{ij} = \frac{1}{2}(\frac{\partial u_i}{\partial x_j} + \frac{\partial u_j}{\partial x_i}), \tag{7.28}$$

where one applies the engineering shear strain $\epsilon_{xy} = 2\varepsilon_{xy}$. Hooke's elasticity can be expressed as

$$\boldsymbol{\sigma} = \mathbf{D}\boldsymbol{\epsilon}, \tag{7.29}$$

where $\mathbf{D}$ is a 6×6 symmetrical matrix as functions of Young's modulus E and Poisson's ratio ν.

Two special cases that are commonly found in many applications are the plane stress ($\sigma_{zz} = 0$, but $\epsilon_{zz} \neq 0$) and plane strain ($\epsilon_{zz} = 0$, but $\sigma_{zz} \neq 0$). The commonly used formulation is the displacement-based formulation or u-based formulation. In the 2-D case, the displacement $\mathbf{u} = (u, v)^T$ and the strain ϵ and stress σ are defined as

$$\boldsymbol{\sigma} = \begin{pmatrix} \sigma_x & \sigma_y & \tau_{xy} \end{pmatrix}^T, \quad \boldsymbol{\epsilon} = \begin{pmatrix} \epsilon_x & \epsilon_y & \gamma_{xy} \end{pmatrix}^T. \tag{7.30}$$

Now the stress-strain relationship becomes

$$\boldsymbol{\sigma} = \mathbf{D}(\boldsymbol{\epsilon} - \boldsymbol{\epsilon_0}), \tag{7.31}$$

where $\mathbf{D}$ is a 3×3 matrix. The strains are given by

$$\epsilon_x = \frac{\partial u}{\partial x}, \quad \epsilon_y = \frac{\partial v}{\partial y}, \quad \gamma_{xy} = \frac{\partial u}{\partial y} + \frac{\partial v}{\partial x}, \tag{7.32}$$

where ϵ_0 is the initial strain due to temperature change or thermal loading. If there is no such change, then the initial strain can be taken to be zero in most applications.

The equilibrium of force in elasticity leads to

$$\nabla \cdot \boldsymbol{\sigma} + \mathbf{b} = \mathbf{0}. \tag{7.33}$$

where $\mathbf{b} = [f_x \quad f_y]^T$ is the body force. In the case of plane stress, we have

$$\mathbf{D} = \frac{E}{1-\nu^2} \begin{pmatrix} 1 & \nu & 0 \\ \nu & 1 & 0 \\ 0 & 0 & (1-\nu)/2 \end{pmatrix}. \tag{7.34}$$

In the case of plane strain, we have

$$\mathbf{D} = \frac{E}{(1-2\nu)} \begin{pmatrix} \frac{1-\nu}{1+\nu} & \frac{\nu}{1+\nu} & 0 \\ \frac{\nu}{1+\nu} & \frac{1-\nu}{1+\nu} & 0 \\ 0 & 0 & \frac{(1-2\nu)}{2(1+\nu)} \end{pmatrix}. \tag{7.35}$$

Clearly, for the 1-D case plane stress when $v = 0, f_y = 0$ and $\sigma_y = \tau_{xy} = 0$, the equation of force balance simply becomes

$$\frac{E}{1-\nu^2} \frac{\partial^2 u}{\partial x^2} + f_x = 0, \tag{7.36}$$

where we have used the stress-strain relationship. This 1-D equation is essentially the same as the 1-D heat transfer equation $u'' + Q = 0$ that will be discussed in detail later, so the solution technique for the 1-D heat transfer shall equally apply. Therefore, we shall focus on the 2-D case in the rest of this section.

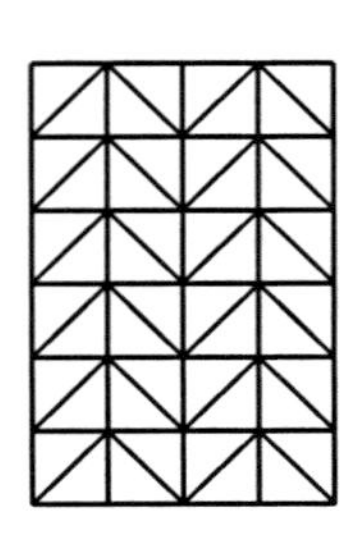

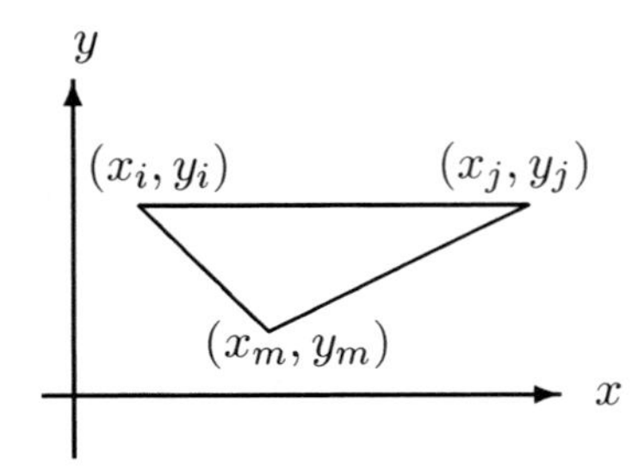

Figure 7.1: Schematic triangular mesh and the layout of a triangular element.

7.2.1 Triangular Elements

Displacements (u, v) in a plane element can be interpolated from nodal displacements. For example, using a triangular element (i, j, m) with three nodal points as shown in Figure 7.1 (x_i, y_i), (x_j, y_j), and (x_m, y_m), we can have relatively simple formulations.

The earliest and simplest triangular element is the constant strain element shown in Fig. 7.2(a), and the displacement field can be expressed as

$$u = \alpha_0 + \alpha_1 x + \alpha_2 y, \tag{7.37}$$

and

$$v = \beta_0 + \beta_1 x + \beta_2 y. \tag{7.38}$$

This gives a constant strain field using equation (7.32)

$$\epsilon_x = \alpha_1, \qquad \epsilon_y = \beta_2, \qquad \gamma_{xy} = \alpha_2 + \beta_1. \tag{7.39}$$

For a linear strain triangular element shown in Fig. 7.2(b), three midpoints are added (marked with $\circ$) because there are six coefficients to be determined. We can express the displacement field as

$$u = \alpha_0 + \alpha_1 x + \alpha_2 y + \alpha_3 x^2 + \alpha_4 xy + \alpha_5 y^2, \tag{7.40}$$

and

$$v = \beta_0 + b_1 x + \beta_2 y + \beta_3 x^2 + \beta_4 xy + \beta_5 y^2. \tag{7.41}$$

Similarly, equation (7.32) gives a strain field

$$\epsilon_x = \alpha_1 + 2\alpha_3 + \alpha_4 y, \qquad \epsilon_y = \beta_2 + \beta_4 x + 2\beta_5 y, \tag{7.42}$$

and

$$\gamma_{xy} = (\alpha_1 + \beta_1) + (\alpha_4 + 2\beta_3)x + (2\alpha_5 + \beta_4)y. \tag{7.43}$$

Obviously, this strain field is linear. In principle, we can design various types of elements with higher order approximations

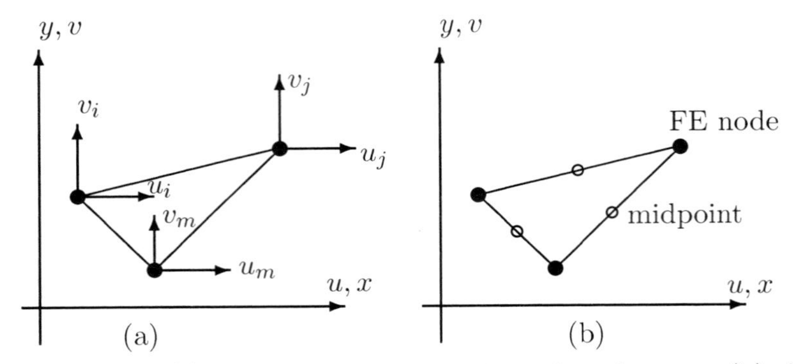

Figure 7.2: (a) A constant strain triangular element; (b) A linear strain triangular element.

and thus higher order accuracy, but they usually require more careful formulations and also more computationally extensive.

For constant strain triangular elements, we have

$$\mathbf{u} = \begin{pmatrix} u \\ v \end{pmatrix} = [N_i\mathbf{I}, N_j\mathbf{I}, N_m\mathbf{I}] \begin{pmatrix} u_i \\ v_i \\ u_j \\ v_j \\ u_m \\ v_m \end{pmatrix} = \mathbf{Nd}, \qquad (7.44)$$

where $\mathbf{I}$ is a 2×2 unitary matrix, i.e.,

$$\mathbf{I} = \begin{pmatrix} 1 & 0 \\ 0 & 1 \end{pmatrix}, \qquad (7.45)$$

and

$$\mathbf{N} = \begin{pmatrix} N_i & 0 & N_j & 0 & N_m & 0 \\ 0 & N_i & 0 & N_j & 0 & N_m \end{pmatrix}. \qquad (7.46)$$

By defining a differential operator

$$\mathbf{L_d} = \begin{pmatrix} \frac{\partial}{\partial x} & 0 \\ 0 & \frac{\partial}{\partial y} \\ \frac{\partial}{\partial y} & \frac{\partial}{\partial x} \end{pmatrix}, \qquad (7.47)$$

we can rewrite the above formulation as

$$\epsilon = \mathbf{L_d u} = \mathbf{L_d N d} = \mathbf{Bd}, \tag{7.48}$$

where

$$\mathbf{B} = \mathbf{L_d N}. \tag{7.49}$$

Now the equation (7.33) becomes

$$\mathbf{Ku} = \mathbf{f}, \tag{7.50}$$

where

$$\mathbf{K} = \int_\Omega \mathbf{B^T D B} dV, \tag{7.51}$$

and

$$f_i = \int_\Omega \mathbf{b} N_i dV + \int_\Gamma \tau N_i d\Gamma, \tag{7.52}$$

where τ is the surface traction (force per unit area) resulting from suitable boundary conditions.

7.2.2 Implementation

In the case of 2-D elastic problems, the simplest elements are linear triangular elements, thus we have $\mathbf{B}_i = \mathbf{L}_d N_i$, or

$$\mathbf{B} = \mathbf{L}_d \mathbf{N} = \frac{1}{2\Delta} \begin{pmatrix} b_i & 0 & b_j & 0 & b_m & 0 \\ 0 & c_i & 0 & c_j & 0 & c_m \\ c_i & b_i & c_j & b_j & c_m & b_m \end{pmatrix}, \tag{7.53}$$

where

$$\Delta = \frac{1}{2} \begin{vmatrix} 1 & x_i & y_i \\ 1 & x_j & y_j \\ 1 & x_m & y_m \end{vmatrix} \tag{7.54}$$

is the area of the triangular element. The linear element implies that the strain is constant in the element. The stiffness matrix $\mathbf{K}_{ij}^{(e)} (i, j = 1, 2, ..., 6)$ of each triangular element can be expressed as

$$\mathbf{K}^{(e)} = \int_{\Omega_e} \mathbf{B}^T \mathbf{D} \mathbf{B} \; dx dy, \tag{7.55}$$

For a triangular element with three nodes (i, j, m), each node has two degrees of freedom (u_i, v_i), then the stiffness matrix $\mathbf{K}^{(e)}$ for each element is a 6×6 matrix. In general, if each element has r nodes, and each node has n degrees of freedom, then the local stiffness matrix is a $rn \times rn$ matrix. If the region has M nodes in total, then Mn equations are needed for this problem. For the present case $(n = 2)$, we need $2M$ equations for plane stress and plane strain.

In order to calculate the contribution of each element to the overall (global) equation, each node should be identified in some way, and most often in terms of the index matrix. The nodal index matrix for three nodes (i, j, m) can be written as

$$\mathrm{ID}_{\mathrm{node}} = \begin{pmatrix} (i,i) & (i,j) & (i,m) \\ (j,i) & (j,j) & (j,m) \\ (m,i) & (m,j) & (m,m) \end{pmatrix}. \tag{7.56}$$

The nodal index matrix identifies the related nodes in the global node-numbering system. However, the main assembly of the stiffness matrix is about the corresponding equations and the application of the boundary conditions, thus we need to transfer the nodal index matrix to the equation index matrix in terms of the global numbering of equations with two degrees of freedom for each node (e.g., equations $2i-1$ and $2i$ for nodal i, equation $2j - 1$ and $2j$ for node j, etc). Thus, for each entry in the stiffness matrix, say (i, j), in the nodal index matrix, we now have four entries, i.e.,

$$(i,j) \to \begin{pmatrix} (2i-1, 2j-1) & (2i-1, 2j) \\ (2i, 2j-1) & (2i, 2j) \end{pmatrix}, \qquad etc \tag{7.57}$$

The equation index matrix now has 6×6 entries, and each entry is a pair such as $(2i - 1, 2j - 1)$, ... , $(2m, 2m)$, etc. By writing it as two index matrices $(ID = JD^T)$, we now have

$$\mathrm{ID}_{\mathrm{equ}} = \mathrm{JD}_{\mathrm{equ}}^T,$$

$$\mathrm{JD}_{\mathrm{equ}} = \begin{pmatrix} 2i-1 & 2i & 2j-1 & 2j & 2m-1 & 2m \\ 2i-1 & 2i & 2j-1 & 2j & 2m-1 & 2m \\ 2i-1 & 2i & 2j-1 & 2j & 2m-1 & 2m \\ 2i-1 & 2i & 2j-1 & 2j & 2m-1 & 2m \\ 2i-1 & 2i & 2j-1 & 2j & 2m-1 & 2m \\ 2i-1 & 2i & 2j-1 & 2j & 2m-1 & 2m \end{pmatrix} \tag{7.58}$$

So that the contribution of $K_{ij}^{(e)}$ to the global matrix K_{ij} is simply

$$K_{[JD_{equ}(I,J),JD_{equ}(I,J)]} = K_{[JD_{equ}(I,J),JD_{equ}(I,J)]} + K_{(I,J)}^{(e)},$$

$$I, J = 1, 2, ..., 6. \tag{7.59}$$

Similarly, the contribution of the body force and external force can be computed

$$f(l) = f(l) + f^{(e)}(l)$$

where $l = 2i-1, 2i, 2j-1, 2j, 2m-1, 2m$ etc.

For simplicity, we now use a 2-D triangular element with nodes i, j, m and coordinates $(x_i, y_i), (x_j, y_j)$ and (x_m, y_m), we have the linear shape functions

$$N_i = \frac{(a_i + b_i x + c_i y)}{\det \begin{vmatrix} 1 & x_i & y_i \\ 1 & x_j & y_j \\ 1 & x_m & y_m \end{vmatrix}}, \tag{7.60}$$

$$a_i = \det \begin{vmatrix} x_j & y_j \\ x_m & y_m \end{vmatrix}, \tag{7.61}$$

$$b_i = y_j - y_m, \qquad c_i = x_m - x_j. \tag{7.62}$$

As an example, we study the 2-D plane stress problem of a rectangular elastic medium with a unit thickness (see Fig. 7.3) where the rectangular region is meshed with triangular elements. This 2-D plane stress problem with a triangular mesh can be solved using the following Matlab program

Figure 7.3: A rectangular beam under plane stress bending.

```
% ------------------------------------------------
% Solving a linear elastic beam bending problem
% by using the finite element method
% ------------------------------------------------
% Program by X S Yang (Cambridge University)
% ------------------------------------------------
% After launching Matlab, please type>elasticity;
% Usage: >elasticity(ni, nj);
%   e.g. >elasticity(20,10);
% ------------------------------------------------

function elasticity(ni,nj)

% ------------------------------------------------
% check number of inputs or use default values
if nargin<2,
   disp('Usage: elasticity(ni,nj)');
   disp(' e.g., elasticity(15,4)');
end

% Default values for ni=20, nj=10 --------------
% ni=number of division along beam axis ---------
% nj=number of division in transverse direction -
if nargin<1, ni=20; nj=10; end

% ------------------------------------------------
% ----- Preprocessing  (build an FE model)  -----
% ------------------------------------------------
% Initializing the parameters and beam geometry
```

```
% Triangular mesh/elements for 2-D plane stress
% Forces at the top end in units of N (Newton)
fx=1000; fy=50;
% Applied/fixed displacements
u0=0.0; v0=0.0;

% Size of the rectangular beam
% tunit=thickness
height=1; width=0.2; tunit=0.1;

% number of nodes (n) and
% number of elements (m)
n=ni*nj; m=2*(ni-1)*(nj-1);
dx=width/(nj-1); dy=height/(ni-1);

% Initialize the matrices
A=zeros(2*n,2*n); f=zeros(1,2*n)'; F=f;

% -----------------------------------------------
% Young's modulus Young's modulus =10^8=100 MPa
% This is a very soft material, real materials
% have E=10MPa (rubber) to 1000GPa (diamond)
% We use small Y to get larger displacements
% Poisson ratio nu=0.25
% -----------------------------------------------
Y=10^8; nu=0.25;

% Penalty factor pen for the application of the
% essential boundary conditions
pen=1000*Y;

% Show the results by multiplying these factors
% =1(true displacement); >1 (exaggerated display)
ushow=1; vshow=1;

% -----------------------------------------------
% Show the mesh and grid
subplot(1,3,1);
pcolor(zeros(ni,nj));
set(gca,'fontsize',14,'fontweight','bold');
title('Mesh Grid');
```

```
% ------------------------------------------------
% Dividing the whole domain into small blocks
% ------------------------------------------------
% Generating nodes on a rectangular region
k=0;
for i=1:ni,
   for j=1:nj,
      k=k+1;
      x(k)=(j-1)*dx; y(k)=(i-1)*dy;
      node(k)=k;
      xp(i,j)=x(k); yp(i,j)=y(k);
   end
end

% ------------------------------------------------
% Elements connectivity E(1,:), E(2,:), E(3,:)
k=-1; nk=0;
bcsize=2*(nj+ni-2);

for i=1:ni-1,
      for j=1:nj-1,
      k=k+2;
      nk=node((i-1)*nj+j);
      E(1,k)=nk;
      E(2,k)=nk+1;
      E(3,k)=nj+nk;
      E(1,k+1)=nk+1;
      E(2,k+1)=nj+nk+1;
      E(3,k+1)=nj+nk;
      end
end
% ------------------------------------------------
% Preparing boundary conditions and loads
% Boundary conditions at the bottom (Fixed)
BC(1:nj)=1:nj;
TOPFORCE(1)=[(ni-1)*nj+1];

% ------------------------------------------------
% Assembly to Ku=f (or Au =f in this program)
% Element-by-element assembly

area=height*width/m; s=1/(2*area);
```

```
G=Y*[1 -nu 0;-nu 1 0;0 0 (1-nu)/2];
for i=1:m,
   ja=node(E(1,i)); x1=x(ja); y1=y(ja);
   jb=node(E(2,i)); x2=x(jb); y2=y(jb);
   jc=node(E(3,i)); x3=x(jc); y3=y(jc);

%Index matrix
%ID=[i1,i1......i1; i2,i2....i2;
%      j1....j1;j2....j2;k1...k1;k2...k2];
%JD=ID'; A(i,j)=A(ID,JD);
a=...
[2*(ja-1)+1 2*ja 2*(jb-1)+1 2*jb 2*(jc-1)+1 2*jc];
aj=[a;a;a;a;a;a]; ai=aj';
B=s*[y2-y3 0  y3-y1 0   y1-y2  0;
      0  x3-x2  0  x1-x3  0   x2-x1;
      x3-x2 y2-y3 x1-x3 y3-y1 x2-x1 y1-y2];
% -----------------------------------------------
% Material properties E,\nu are included in G,K
% Stiffness matrix K
   K=B'*G*B*tunit*area;

   for ii=1:6,
      for jj=1:6,
        A(ai(ii,jj),aj(ii,jj))=...
           K(ii,jj)+A(ai(ii,jj),aj(ii,jj));
      end
   end
end

% -----------------------------------------------
% Application of boundary conditions and loads
% -----------------------------------------------
% Essential boundary conditions (u_i=\bar u_i)
for i=1:size(BC,2),
      ij=2*BC(i);
      A(ij-1,ij-1)=A(ij-1,ij-1)+pen;
      F(ij-1)=f(ij-1);
      f(ij-1)=f(ij-1)+pen*u0;
      A(ij,ij)=A(ij,ij)+pen;
      F(ij)=f(ij); f(ij)=f(ij)+pen*v0;
      text(i-0.1,1,'\Delta');
end
```

```
% ----- Application of loads --------------------
for i=1:size(TOPFORCE,2),
   ij=2*TOPFORCE(i);
   f(ij-1)=f(ij-1)+fx;
   F(ij-1)=f(ij-1);
   f(ij)=f(ij)+fy;
   F(ij)=f(ij);
   text(0.5,ni,'\rightarrow',...
          'fontsize',20,'color','r');
end
% ------------End of Preprocessing -------------

% ---------------------------------------------
% ----- Solving the equation/problem ------------

% Solving the equation by simple inversion
u=A\f;

% ---------------------------------------------
% ----- Postprocessing and visualisation --------
% ---------------------------------------------

% Preparing and decomposing the results
k=-1;
for i=1:ni,
   for j=1:nj,
      k=k+2;
      U(i,j)=u(k);
      V(i,j)=u(k+1);
      Fx(i,j)=F(k);
      Fy(i,j)=F(k+1);
   end
end

% ---------------------------------------------
% Visualising the displacements using pcolor() --

subplot(1,3,2);
pcolor(xp+U*ushow,yp+V*vshow,U);
shading interp; colorbar;
set(gca,'fontsize',14,'fontweight','bold');
```

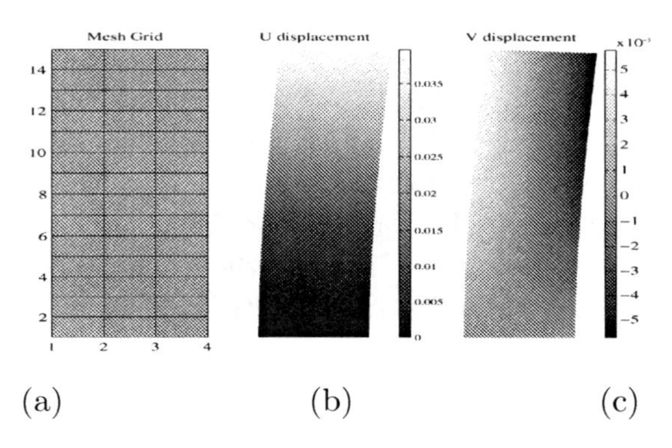

(a) (b) (c)

Figure 7.4: Beam bending a) initial grid, b) horizontal displacement (U), c) vertical displacement (V).

```
axis off;
title('U displacement');
subplot(1,3,3);
pcolor(xp+U*ushow,yp+V*vshow,V);
axis off;
shading interp;
set(gca,'fontsize',14,'fontweight','bold');
colorbar;
title('V displacement');
% -------End of program -------------------------
% ----------------------------------------------
```

Using the above program, we can simulate a 2-D plane stress problem for a rectangular beam with one end fixed and a unit force applied at the top (see Fig. 7.3).

Figure 7.4 shows the initial grid, the horizontal and vertical displacement of the beam under a concentrated loading at the top. This program can be easily modified to simulate all sorts of elasticity problems by changing the boundary conditions, material properties (E, ν) and the geometry of the beam.

From this simple program of simulating the elastic deformation of a beam, we know that the basic procedure of the finite element analysis includes three parts: preprocessing, solving the matrix equations, and postprocessing. In the preprocessing step, we build a model including dividing the whole domain into many small blocks with nodes and elements. We also assign material properties such as E, ρ, ν and apply the boundary conditions and loads. The main objective of this step is to produce a unique $\mathbf{K}\mathbf{u} = \mathbf{f}$. In the solving step, we try to invert the equation to get the solution $\mathbf{u}$ or the displacements. The final postprocessing step is to visualise the results using coloured contours or to calculate other quantities such as stresses and strains from the displacements. In fact, all the finite element packages should include these three major components. We will discuss finite element packages in detail in the final chapter.

Chapter 8

Heat Conduction

Heat transfer problems are very common in engineering and computational modelling. The geometry in most applications is irregular. Thus, finite element methods are especially useful in this case.

8.1 Basic Formulation

The steady-state heat transfer is governed by the heat conduction equation or Poisson's equation

$$\nabla \cdot (k \nabla u) + Q = 0, \tag{8.1}$$

with the essential boundary condition

$$u = \bar{u}, \qquad \mathbf{x} \in \partial\Omega_E, \tag{8.2}$$

and the natural boundary condition

$$k \frac{\partial u}{\partial n} - q = 0, \qquad \mathbf{x} \in \partial\Omega_N. \tag{8.3}$$

Multiplying both sides of equation (8.1) by the shape function N_i and using the formulation similar to the formulation (6.36) in terms of $u \approx u_h$, we have

$$\int_\Omega [\nabla \cdot (k \nabla u) + Q] N_i d\Omega - \int_{\partial\Omega_N} [k \frac{\partial u}{\partial n} - q] N_i d\Gamma = 0. \tag{8.4}$$

Integrating by parts and using Green's theorem, we have

$$-\int_\Omega (\nabla u_h \cdot k \cdot \nabla N_i) d\Omega + \int_{\partial\Omega} k \frac{\partial u_h}{\partial n} N_i d\Gamma$$

$$+ \int_\Omega Q N_i d\Omega - \int_{\partial\Omega_N} [k \frac{\partial u_h}{\partial n} - q] N_i d\Gamma = 0. \qquad (8.5)$$

Since $N_i = 0$ on $\partial\Omega_E$, thus we have

$$\int_{\partial\Omega} [\;] N_i d\Gamma = \int_{\partial\Omega_N} [\;] N_i d\Gamma. \qquad (8.6)$$

Therefore, the above weak formulation becomes

$$\int_\Omega (\nabla u_h \cdot k \cdot \nabla N_i) d\Omega - \int_\Omega Q N_i d\Omega - \int_{\partial\Omega_N} q N_i d\Gamma = 0. \quad (8.7)$$

Substituting $u_h = \sum_{j=1}^M u_j N_j(\mathbf{x})$ into the equation, we have

$$\sum_{j=1}^M [\int_\Omega (k \nabla N_i \cdot \nabla N_j) d\Omega] u_j - \int_\Omega Q N_i d\Omega$$

$$- \int_{\partial\Omega_N} q N_i d\Gamma = 0. \qquad (8.8)$$

This can be written in the compact matrix form

$$\sum_{j=1}^M K_{ij} U_j = f_i, \qquad \mathbf{KU} = \mathbf{f}, \qquad (8.9)$$

where $\mathbf{K} = [K_{ij}], (i, j = 1, 2, ..., M)$, $\mathbf{U}^T = (u_1, u_2, ..., u_M)$, and $\mathbf{f}^T = (f_1, f_2, ..., f_M)$. That is,

$$K_{ij} = \int_\Omega k \nabla N_i \nabla N_j d\Omega, \qquad (8.10)$$

$$f_i = \int_\Omega Q N_i d\Omega + \int_{\partial\Omega_N} q N_i d\Gamma. \qquad (8.11)$$

◇ **Example 8.1:** As a simple example, we consider the 1-D steady-state heat conduction problem,

$$u''(x) + Q(x) = 0,$$

with boundary conditions

$$u(0) = \beta, \qquad u'(1) = q.$$

For a special case $Q(x) = r\exp(-x)$, we have the analytical solution

$$u(x) = (\beta - r) + (re^{-1} + q)x + re^{-x}. \tag{8.12}$$

Then equation (8.11) becomes

$$\sum_{j=1}^{M}\Big(\int_0^1 N_i' N_j' dx\Big)u_j = \int_0^1 QN_i dx + qN_i(1).$$

For the purpose of demonstrating the implementation procedure, let us solve this problem by dividing the interval into 4 elements and 5 nodes. This will be discussed later in more details. ◇

8.2 Element-by-Element Assembly

The assembly of the linear matrix system is the popular element-by-element method. The stiffness matrix $\mathbf{K}$ in equations (8.9) and (8.11) is the summation of the integral over the whole solution domain, and the domain is now divided into m elements with each element on a subdomain Ω_e $(e = 1, 2, ..., m)$. Each element contributes to the whole stiffness matrix, and in fact, its contribution is a pure number. Thus, assembly of the stiffness matrix can be done in an element-by-element manner. Furthermore, $K_{i,j} \neq 0$ if and only if (or *iff*) nodes i and j belong to the same elements. In the 1-D case, $K_{i,j} \neq 0$ only for $j = i-1, i, i+1$. In finite element analysis, the shape functions N_j are typically localized functions, thus the matrix $\mathbf{K}$ is usually sparse in most cases.

The element-by-element formulation can be written as

$$K_{i,j} = \sum_{e=1}^{m} K_{i,j}^{(e)}, \qquad K_{i,j}^{(e)} = \int_{\Omega_e} k\nabla N_i \nabla N_j d\Omega_e, \tag{8.13}$$

and

$$f_i = \sum_{e=1}^{m} f_i^{(e)}, \qquad f_i^{(e)} = \int_{\Omega_e} QN_i d\Omega_e + \int_{\partial\Omega_{N_e}} qN_i d\Gamma_e. \tag{8.14}$$

In addition, since the contribution of each element is a simple number, the integration of each element can be done using the local coordinates and local node numbers or any coordinate system for the convenience of integration over an element. Then, the nonzero contribution of each element to the global system matrix **K** is simply assembled by direct addition to the corresponding global entry (of the stiffness matrix) of the corresponding nodes or related equations. In reality, this can be easily done using an index matrix to trace the element contribution to the global system matrix.

◇ **Example 8.2:** The assembly of the global system matrix for the example with 4 elements and five nodes is shown below. For each element with i and j nodes, we have

$$N_i = 1 - \xi, \qquad N_j = \xi, \qquad \xi = \frac{x}{L}, \qquad L = h_e,$$

$$K_{ij}^{(e)} = [\int_0^L k N_i' N_j' dx] = \frac{k}{h_e}\begin{pmatrix} 1 & -1 \\ -1 & 1 \end{pmatrix},$$

$$f_i^{(e)} = \frac{Qh_e}{2}\begin{pmatrix} 1 \\ 1 \end{pmatrix},$$

so that, for example in elements 1 and 2, these can extend to all nodes (with $h_i = x_{i+1} - x_i, i = 1, 2, 3, 4$),

$$K^{(1)} = \begin{pmatrix} k/h_1 & -k/h_1 & 0 & 0 & 0 \\ -k/h_1 & k/h_1 & 0 & 0 & 0 \\ 0 & 0 & 0 & 0 & 0 \\ 0 & 0 & 0 & 0 & 0 \\ 0 & 0 & 0 & 0 & 0 \end{pmatrix}, \quad f^{(1)} = \frac{Q}{2}\begin{pmatrix} h_1 \\ h_1 \\ 0 \\ 0 \\ 0 \end{pmatrix},$$

$$K^{(2)} = \begin{pmatrix} 0 & 0 & 0 & 0 & 0 \\ 0 & k/h_2 & -k/h_2 & 0 & 0 \\ 0 & -k/h_2 & k/h_2 & 0 & 0 \\ 0 & 0 & 0 & 0 & 0 \\ 0 & 0 & 0 & 0 & 0 \end{pmatrix}, \quad f^{(2)} = \frac{Q}{2}\begin{pmatrix} 0 \\ h_2 \\ h_2 \\ 0 \\ 0 \end{pmatrix},$$

and so on. Now the global system matrix becomes

$$K = \begin{pmatrix} k/h_1 & -k/h_1 & 0 & 0 & 0 \\ -k/h_1 & \frac{k}{h_1} + \frac{k}{h_2} & -k/h_2 & 0 & 0 \\ 0 & -k/h_2 & \frac{k}{h_2} + \frac{k}{h_3} & -k/h_3 & 0 \\ 0 & 0 & -k/h_3 & \frac{k}{h_3} + \frac{k}{h_4} & -k/h_4 \\ 0 & 0 & 0 & -k/h_4 & k/h_4 \end{pmatrix},$$

$$\mathbf{U} = \begin{pmatrix} u_1 \\ u_2 \\ u_3 \\ u_4 \\ u_5 \end{pmatrix}, \qquad \boldsymbol{f} = \begin{pmatrix} Qh_1/2 \\ Q(h_1 + h_2)/2 \\ Q(h_2 + h_3)/2 \\ Q(h_3 + h_4)/2 \\ Qh_4/2 + q \end{pmatrix},$$

where the last row of $\boldsymbol{f}$ has already included the natural boundary condition at $u'(1) = q$. ◇

8.3 Application of Boundary Conditions

Boundary conditions can be essential, natural or mixed. The essential boundary conditions are automatically satisfied in the finite element formulation by the approximate solution. These include the displacement, rotation, and known value of the solution. Sometimes, they are also called the geometric boundary conditions. In our example, it is $u(0) = \beta$. Natural boundary conditions often involve the first derivatives such as strains, heat flux, force, and moment. Thus, they are also referred to as force boundary conditions. In our example, it is $u'(1) = q$.

The natural boundary conditions are included in the integration in the finite element equations such as (8.11). Thus no further imposition is necessary. On the other hand, although the essential boundary conditions are automatically satisfied in the finite element formulations, they still need to be implemented in the assembled finite element equations to ensure unique solutions. The imposition of the essential boundary conditions can be done in main several ways: a) direct application; b) Lagrangian multiplier and c) penalty method. To show how these methods work, we use the 1-D poisson equation on the distinct points $x_i (i = 1, 2, ..., M) \in [0, 1]$ to aid our discussion.

8.3.1 Direct Application

In this method, we simply use the expansion $u_h = \sum_{i=1}^{M} u_i N_i$, and apply directly the essential boundary conditions at point i to replace the corresponding ith equation with $u_i = \bar{u}_i$ so that ith row of the stiffness matrix $\mathbf{K}$ in equation (8.9) becomes $(0, 0, ..., 1, ..., 0)$ and the corresponding $f_i = f(i) = \bar{u}_i$. All other points will be done in the similar manner. For example, the boundary conditions $u(0) = \alpha$ and $u(M) = \beta$ in the 1-D case mean that the first and last equations are replaced by $u_1 = \alpha$ and $u_M = \beta$, respectively. Thus, $K_{11} = 1, f_1 = \alpha$ (all other coefficients are set to be zeros: $K_{12} = ... = K_{1M} = 0$, and $K_{MM} = 1, f_M = \beta$ with $K_{M1} = ... = K_{M,M-1} = 0$). Then, the equations can be solved for $(u_1, u_2, ..., u_M)^T$. This method is widely used due to its simplicity and the advantage of time-stepping because it allows bigger time steps.

8.3.2 Lagrangian Multiplier

This method is often used in the structure and solid mechanics to enforce the constraints $(u_i = \bar{u}_i)$. The variation is added by the extra term $\lambda(u_i - \bar{u}_i)$ where λ is the Lagrange multiplier. Now we have

$$\Pi = \frac{1}{2}\mathbf{u^T K u} - \mathbf{u^T f} + \lambda(u_i - \bar{u}_i), \tag{8.15}$$

whose variation $\delta\Pi = 0$ leads to

$$\delta\mathbf{u^T K u} - \delta\mathbf{u^T f} + \lambda\delta u_i + \delta\lambda(u_i - \bar{u}_i) = 0. \tag{8.16}$$

Because $\delta\mathbf{u}$ and $\delta\lambda$ are arbitrary, we have

$$\begin{pmatrix} \mathbf{K} & \boldsymbol{e}_i \\ \boldsymbol{e}_i^T & 0 \end{pmatrix} \begin{pmatrix} \mathbf{u} \\ \lambda \end{pmatrix} = \begin{pmatrix} \mathbf{f} \\ \bar{u}_i \end{pmatrix}.$$

where $\boldsymbol{e}_i = (0, 0, ..., 1, 0, ..., 0)^T$ (its ith entry is equal to one). This method can be extended to m Lagrangian multipliers.

8.3.3 Penalty Method

One of the most widely used methods of enforcing the essential boundary conditions is the so-called penalty method in terms of a very large coefficient γ, $u_i = \bar{u}$ at $\mathbf{x}_i \in \partial\Omega_E$, so that $\gamma u_i = \gamma\bar{u}$ can be directly added onto $\mathbf{Ku} = \mathbf{f}$. In the 1-D example, it simply leads to $K_{11} = K_{11}+\gamma, K_{MM} = K_{MM}+\gamma$, and $f_1 = f_1 + \gamma\alpha, f_M = f_M + \gamma\beta$. The common rule for choosing γ is that $\gamma \gg \max|K_{ii}|$. Usually, $\gamma \approx 1000 \max|K_{ii}|$ should be adequate. The penalty method is widely used in steady-state problems. However, it may affect the efficiency of time-stepping since it increases the maximum eigenvalue of the stiffness matrix, and thus very small time steps are required for convergence. The advantage of the penalty method is that the handling of the essential boundary conditions becomes simpler from the implementation point of view. The disadvantage is that the conditions are only satisfied approximately.

◇ **Example 8.3:** Following the same example of the 1-D steady state heat conduction discussed earlier, we now use the direct application method for the essential boundary conditions. We can replace the first equation $\sum_{j=1}^{5} K_{1j}u_j = f_1$ with $u_1 = \beta$, so that the first row becomes $K_{1j} = (1\,0\,0\,0\,0)$ and $f_1 = \beta$. Thus, we have

$$\mathbf{K} = \begin{pmatrix} 1 & 0 & 0 & 0 & 0 \\ -k/h_1 & \frac{k}{h_1}+\frac{k}{h_2} & -k/h_2 & 0 & 0 \\ 0 & -k/h_2 & \frac{k}{h_2}+\frac{k}{h_3} & -k/h_3 & 0 \\ 0 & 0 & -k/h_3 & \frac{k}{h_3}+\frac{k}{h_4} & -k/h_4 \\ 0 & 0 & 0 & -k/h_4 & k/h_4 \end{pmatrix},$$

$$\mathbf{U} = \begin{pmatrix} u_1 \\ u_2 \\ u_3 \\ u_4 \\ u_5 \end{pmatrix}, \qquad \boldsymbol{f} = \begin{pmatrix} \beta \\ Q(h_1+h_2)/2 \\ Q(h_2+h_3)/2 \\ Q(h_3+h_4)/2 \\ Qh_4/2+q \end{pmatrix},$$

For the case of $k = 1, Q = -1, h_1 = ... = h_4 = 0.25$, $\beta = 1$ and

$q = -0.25$, we have

$$\mathbf{K} = \begin{pmatrix} 1 & 0 & 0 & 0 & 0 \\ -4 & 8 & -4 & 0 & 0 \\ 0 & -4 & 8 & -4 & 0 \\ 0 & 0 & -4 & 8 & -4 \\ 0 & 0 & 0 & -4 & 4 \end{pmatrix}, \qquad \boldsymbol{f} = \begin{pmatrix} 1 \\ -0.25 \\ -0.25 \\ -0.25 \\ -0.375 \end{pmatrix}$$

Hence, the solution is

$$\mathbf{U} = \mathbf{K}^{-1}\boldsymbol{f} = \begin{pmatrix} 1.00 \\ 0.72 \\ 0.50 \\ 0.34 \\ 0.25 \end{pmatrix}.$$

Chapter 9

Transient Problems

9.1 The Time Dimension

The problems we have discussed so far are static or time-independent because the time dimension is not involved. For time-dependent problems, the standard finite element formulation first produces an ordinary differential equation for matrices rather than algebraic matrix equations. Therefore, besides the standard finite element formulations, extra time-stepping schemes should be used in a similar manner as that in finite difference methods.

As the weak formulation uses the Green theorem that involves the spatial derivatives, the time derivatives can be considered as the source term. Thus, one simple and yet instructive way to extend the finite element formulation to include the time dimension is to replace Q in equation (8.1) with $Q - \alpha u_t - \beta u_{tt} = Q - \alpha \dot{u} - \beta \ddot{u}$ so that we have

$$\nabla \cdot (k \nabla u) + (Q - \alpha \dot{u} - \beta \ddot{u}) = 0. \qquad (9.1)$$

The boundary conditions and initial conditions are $u(\mathbf{x}, 0) = \phi(\mathbf{x})$, $u = \mathbf{u}, \mathbf{x} \in \partial\Omega_E$, and $k\frac{\partial u}{\partial n} - q = 0, \mathbf{x} \in \partial\Omega_N$. Using integration by parts and the expansion $u_h = \sum_{j=1}^{M} u_j N_j$, we have

$$\sum_{j=1}^{M} [\int_{\Omega} (k \nabla N_i \nabla N_j) d\Omega]$$

$$+\sum_{j=1}^{M}\int_{\Omega}[(N_i\alpha N_j)\dot{u}_j+(N_i\beta N_j)\ddot{u}_j]d\Omega$$

$$-\int_{\Omega}N_iQd\Omega-\int_{\partial\Omega_N}N_iqd\Gamma=0, \tag{9.2}$$

which can be written in a compact form as

$$\mathbf{M}\ddot{\mathbf{u}}+\mathbf{C}\dot{\mathbf{u}}+\mathbf{K}\mathbf{u}=\mathbf{f}, \tag{9.3}$$

where

$$K_{ij}=\int_{\Omega}[(k\nabla N_i\nabla N_j)]d\Omega, \tag{9.4}$$

$$f_i=\int_{\Omega}N_iQd\Omega+\int_{\partial\Omega_N}N_iqd\Gamma, \tag{9.5}$$

and

$$C_{ij}=\int_{\Omega}N_i\alpha N_jd\Omega, \qquad M_{ij}=\int_{\Omega}N_i\beta N_jd\Omega. \tag{9.6}$$

The matrices $\mathbf{K},\mathbf{M},\mathbf{C}$ are symmetric, that is to say, $K_{ij}=K_{ji}, M_{ij}=M_{ji}, C_{ij}=C_{ji}$ due to the interchangeability of the orders in the product of the integrand k, N_i and N_j (i.e., $\nabla N_i\cdot k\cdot\nabla N_j=k\nabla N_i\nabla N_j$, $N_i\alpha N_j=N_j\alpha N_i=\alpha N_iN_j$ etc). The matrix $\mathbf{C}=[C_{ij}]$ is the damping matrix similar to the damping coefficient of damped oscillations. $\mathbf{M}=[M_{ij}]$ is the general mass matrix due to a similar role acting as an equivalent mass in dynamics. In addition, before the boundary conditions are imposed, the matrix is usually singular, which may imply many solutions. Only after the proper boundary conditions have been enforced, the stiffness matrix will be nonsingular, thus unique solutions may be obtained. On the other hand, $\mathbf{M}$ and $\mathbf{C}$ will be always non-singular if they are not zero. For example, for the 1-D elements (with nodes i and j),

$$K_{ij}^{(e)}=\frac{k}{h_e}\begin{pmatrix}1 & -1\\ -1 & 1\end{pmatrix}, \qquad \det[K^{(e)}]=0, \tag{9.7}$$

but

$$M_{ij}^{(e)}=\frac{\beta h_e}{6}\begin{pmatrix}2 & 1\\ 1 & 2\end{pmatrix}, \qquad \det[M^{(e)}]\neq 0,$$

$$C_{ij}^{(e)} = \frac{\alpha h_e}{6}\begin{pmatrix} 2 & 1 \\ 1 & 2 \end{pmatrix}, \qquad \det[C^{(e)}] \neq 0, \tag{9.8}$$

Clearly, if $\mathbf{M} = 0$, it reduces to the linear heat conduction. If $\mathbf{C} = 0$, it becomes the wave equation with the source term.

9.2 Time-Stepping

From the general governing equation

$$\mathbf{M}\ddot{\mathbf{u}} + \mathbf{C}\dot{\mathbf{u}} + \mathbf{K}\mathbf{u} = \mathbf{f}, \tag{9.9}$$

we see that it is an ordinary differential equation in terms of time and matrices. Thus, in principle, all the time-stepping methods developed in the standard finite difference method can be used for this purpose. For a simple center difference scheme, we have

$$\dot{\mathbf{u}} = \frac{\mathbf{u}^{n+1} - \mathbf{u}^n}{\Delta t}, \qquad \ddot{\mathbf{u}} = \frac{(\mathbf{u}^{n+1} - 2\mathbf{u}^n + \mathbf{u}^{n-1})}{(\Delta t)^2}. \tag{9.10}$$

so that equation (9.9) becomes

$$\mathbf{M}\frac{(\mathbf{u}^{n+1} - 2\mathbf{u}^n + \mathbf{u}^{n-1})}{(\Delta t)^2}$$

$$+\mathbf{C}\frac{(\mathbf{u}^{n+1} - \mathbf{u}^{n-1})}{2\Delta t} + \mathbf{K}\mathbf{u}^n = \mathbf{f}. \tag{9.11}$$

Now the aim is to express $\mathbf{u}^{n+1}$ in terms of $\mathbf{u}^n$ and $\mathbf{u}^{n-1}$.

9.2.1 1-D Transient Heat Transfer

In the case of heat conduction ($\mathbf{M} = 0$), we have

$$\mathbf{C}\dot{\mathbf{u}} + \mathbf{K}\mathbf{u} = \mathbf{f}, \tag{9.12}$$

or

$$\dot{\mathbf{u}} = \mathbf{C}^{-1}(\mathbf{f} - \mathbf{K}\mathbf{u}). \tag{9.13}$$

Using the explicit time-stepping method, we can write it as

$$\frac{\mathbf{u}^{n+1} - \mathbf{u}^n}{\Delta t} = \mathbf{C}^{-1}(\mathbf{f} - \mathbf{K}\mathbf{u}^n), \tag{9.14}$$

so that we have

$$\mathbf{u}^{n+1} = \mathbf{u}^n + \Delta t \mathbf{C}^{-1}(\mathbf{f} - \mathbf{K}\mathbf{u}^n). \tag{9.15}$$

◇ **Example 9.1:** For a transient heat conduction problem, we have

$$\alpha u_t = k u_{xx} + Q,$$

and

$$u(x,0) = 0, \;\; u(0,t) = 1, \;\; u'(1) = q.$$

The formulation with 5 nodes and 4 elements leads to

$$\mathbf{C} = \frac{\alpha}{6} \begin{pmatrix} 2h_1 & h_1 & 0 & 0 & 0 \\ h_1 & 2(h_1+h_2) & h_2 & 0 & 0 \\ 0 & h_2 & 2(h_2+h_3) & h_3 & 0 \\ 0 & 0 & h_3 & 2(h_3+h_4) & h_4 \\ 0 & 0 & 0 & h_4 & 2h_4 \end{pmatrix}.$$

For the case of $\alpha = 6, k = 1, Q = -1, h_1 = ... = h_4 = 0.25$, we have

$$\mathbf{C} = \begin{pmatrix} 0.5 & 0.25 & 0 & 0 & 0 \\ 0.25 & 1 & 0.25 & 0 & 0 \\ 0 & 0.25 & 1 & 0.25 & 0 \\ 0 & 0 & 0.25 & 1 & 0.25 \\ 0 & 0 & 0 & 0.25 & 0.5 \end{pmatrix}.$$

◇

9.2.2 Wave Equation

For the wave equation ($\mathbf{C} = 0$), we have

$$\mathbf{M}\ddot{\mathbf{u}} + \mathbf{K}\mathbf{u} = \mathbf{f}. \tag{9.16}$$

Using

$$\ddot{\mathbf{u}} = \frac{\mathbf{u}^{n+1} - 2\mathbf{u}^n + \mathbf{u}^{n-1}}{(\Delta t)^2}, \tag{9.17}$$

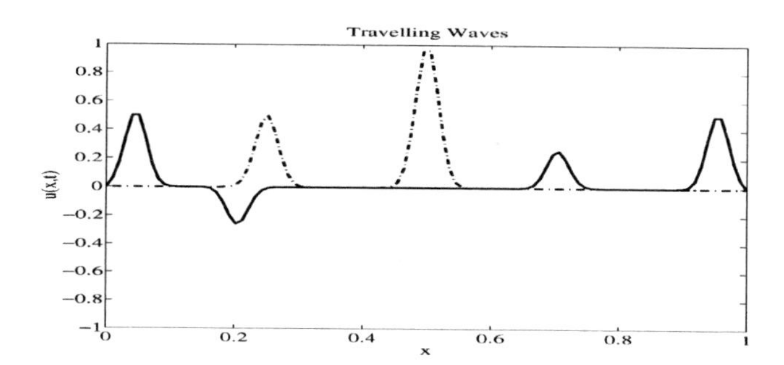

Figure 9.1: 1-D wave equation solved by the FEM.

we have

$$\mathbf{u}^{n+1} = \mathbf{M}^{-1}\mathbf{f}(\Delta t)^2 + [2\mathbf{I} - (\Delta t)^2\mathbf{M}^{-1}\mathbf{K}]\mathbf{u}^n - \mathbf{u}^{n-1}, \quad (9.18)$$

where $\mathbf{I}$ is an identity or unit matrix. For example, the 1-D wave equation

$$\frac{\partial^2 u}{\partial t^2} = c\frac{\partial^2 u}{\partial x^2}, \quad (9.19)$$

with the boundary conditions

$$u(0) = u(1) = 0, \qquad u(x,0) = e^{-(x-1/2)^2}, \quad (9.20)$$

can be written as

$$M_{ij} = \int_0^1 N_i N_j dx, \;\; K_{ij} = \int_0^1 cN_i' N_j' dx, \;\; \mathbf{f} = 0, \quad (9.21)$$

and $\mathbf{u}^0$ is derived from the $u(x,0) = \exp[-(x-1/2)^2]$.

This problem can be solved using the following Matlab program:

```
% ------------------------------------------------
% Solving the 1-D wave equation using the
% finite element method, implemented in Matlab
```

```
% written by X S Yang (Cambridge University)
% PDE form: u_{tt}-c^2 u_{xx}=0;   c=speed=1;
% ------------------------------------------------
%n=number of nodes, N=time-step
n=100;

% ----- Initializing various parameters ---------
L=1.0;               % length of domain
speed=1.0;           % wave speed
m=n-1;               % number of elements
time=1;              % total time of simulations

% ----- Time steps and element size -------------
dt=L/(n*speed);  hh=L/m;
N=time/dt;           % Number of time steps

% ----- Preprocessing ---------------------------
% Split the domain into regularly-spaced n nodes
for i=1:n,
           x(i)=(i-1)*L/m;
end
x(1)=0; x(n)=L;

% ----- Finding the element connectivity --------
% Simple 1-D 2-node elements E(1,:) and E(2,:)
for i=1:m,
   E(1,i)=i;
   E(2,i)=i+1;
   h(i)=abs(x(E(2,i))-x(E(1,i)));
end

% ----- Initialization of arrays/matrices --------
u=zeros(1,n)'; f=zeros(1,n)';
k=[1 -1;-1 1];
A=zeros(n,n);  M=zeros(n,n);

% ----- Element-by-element assembly --------------
% M d^2U/dt^2+KU=f;
for i=1:m,
   A(i,i)=A(i,i)+1/h(i);
   A(i,i+1)=A(i,i+1)-1/h(i);
   A(i+1,i)=A(i+1,i)-1/h(i);
```

```
   A(i+1,i+1)=A(i+1,i+1)+1/h(i);
end

% ----- Application of boundary conditions ------
% Fixed boundary  at both ends: u(0)=u(1)=0
A(n,1)=1; A(n,n-1)=0; A(1,1)=1; A(1,2)=0;

% ----- General mass matrix M ------------------
for i=2:n-1,
    M(i,i)=hh;
end;
M(1,1)=hh/2;
M(n,n)=hh/2; Minv=inv(M);

% ----- Preparing time stepping ----------------
D=2*eye(n,n)-dt*dt*Minv*A;

% ----- Initial waveforms with two peaks --------
u0=exp(-(40*(x-L/2)).^2)...
            +0.5*exp(-(40*(x-L/4)).^2);
v=u0'; U=v;

% ----- Plot out the initial waveforms ----------
plot(x,u0); axis([0 L -1 1]);
set(gca,'fontsize',14,'fontweight','bold');
title('Travelling Waves');
xlabel('x'); ylabel('u(x,t)');
axis tight;
set(gca,'nextplot','replacechildren');

% ---------------------------------------------
% ---- Solving the matrix equation -------------
% ---- Start time-stepping ---------------------
for t=1:N,
        u=D*U-v;
        v=U;          % stored to be used later
        U=u;          % stored as previous values

% ----- Display the travelling wave -------------
% ----- Postprocessing -------------------------
    plot(x,u,x,u0,'-.','linewidth',2);
    axis([0 L -1 1]);
```

```
    V(t)=getframe;    % for animation/movie
end

% ----- Replay the animation --------------------
movie(V);
% ----------------------------------------------
```

The initial wave profile is split into two waves, one travels to the left and one to the right. Figure 9.1 shows a snap shot of the travelling wave at time $t = 1$.

The finite element methods in this book are mainly for linear partial differential equations. Although these methods can in principle be extended to nonlinear problems, however, some degrees of approximations and linearization are needed. In addition, an iterative procedure is required to solve the resultant nonlinear matrix equations. The interested readers can refer to many excellent books on these topics.

All the methods we have discussed in this book are becoming conventional methods. They are very useful and form the basis for further research and studies. In fact, researchers have formulated many new methods. Two of the most promising methods are the discrete element method and element-free or meshless method. The former concerns a system of interacting particles and it can be used to simulate various fluid and granular flows. The latter does not use the mesh at all, it only has isolated nodes and this could have distinct advantages over conventional methods as there is no mesh degeneration in the element-free method.

Finite element analysis is only a part of scientific computing which is a multisciplinary subject, and there are a lot of challenging problems waiting for you to solve. Most importantly, there are many other methods to be invented by you, the readers.

Chapter 10

Finite Element Packages

Once you understand the fundamentals of finite element methods and master the basic skills of computer programming, you can start to do various computer simulations by doing your own programming. This is usually time consuming. There is an easier alternative: that is to use the scientific computing packages, either commercial or free open source softwares. Commercial softwares (for examples, Abaqus, Ansys, Fluent, Femlab) are professionally developed. They are general-purpose packages and thus can do a wide range of simulations and applications, however, they are often expensive. There are dozens of free software available as well, so that you can use to serve your purpose. Some even have source codes and you can add more functions if you have time to do so. Whatever software you use for your finite element analysis (FEA), they usually have three major components: pre-processing, the main solver, and post-processing. You have to go through each of these three stages to build a finite element model (with appropriate material properties, boundary conditions and loads), to conduct the numerical analysis, and to visualize your results. Once you have the results, you have to explain and make sense of these results.

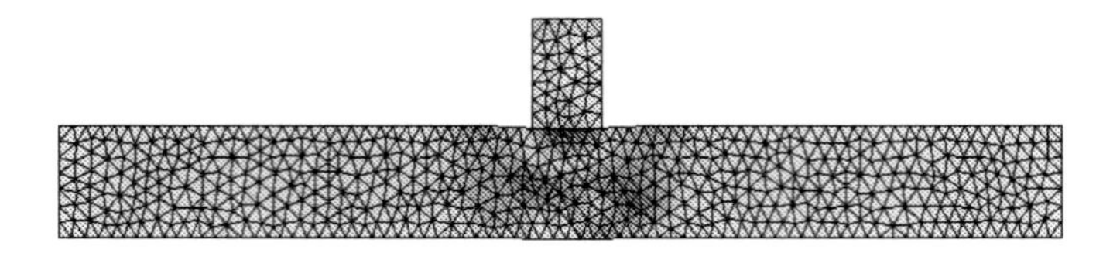

Figure 10.1: Mesh geneation and numerical model for the impact simulation of a drop load on a brittle beam.

10.1 FEA Procedure

10.1.1 Preprocessing

Preprocessing is the step that you prepare or translate your mathematical model of engineering problems into the format and model data that are compatible to the software you are using. This usually includes: 1) setting up the geometry of the problem such as the shape, size, and domain; 2) generating mesh (node, connectivity, element type); 3) assigning material properties such as elasticity, plasticity, and fluid; 4) applying boundary conditions and constraints such initial conditions and various boundary conditions; 5) applying the loads (distributed, concentrated, static, dynamic, or time-dependent load). This will lead to a numerical model in your computer, stored as files. Luckily, most commercial software packages have powerful preprocessing functionalities and the only thing you have to do is to follow their tutorials and demos, or if you have to, read their manuals (usually hundreds or even thousands of pages). Fore example, Figure 10.1 shows the mesh generation and triangular mesh required for a typical simulation of impact problems where a drop load impacts on a brittle plate.

The most time-consuming part is probably to generate the correct mesh and select the correct element types. A sparse mesh with larger elements will result in a small number of degrees of freedom, and thus it is quicker to get the results, however, the error is usually large. A dense mesh with smaller size of elements will give better convergence but could be extremely time-consuming. In theory, when $h \to 0$, a finite element system should converge to a continuum system. However, for problems involving fracture dynamics, this is not always the case. Therefore, there is a tradeoff between the mesh size and total number of degrees of freedom, and choice of the right mesh size depends on the accuracy of your desired solution, the type of problem and computing resources.

The choice of right element types is also very important. A wrong type of elements, such as replacing solid elements with shell elements, could results in incorrect results. Selecting the appropriate element types is crucial for any finite element analysis. If you use simple linear elements, you can never get the accuracy as provided by higher-order elements. In addition, some elements may be easily distorted, and thus resulting inappropriate results. However, most finite element packages provides very good examples and guidance on the choice of element types for a given type of problem, though the choice in most case is not unique. The actual use of a particular element type again depends on the type of problem and the required accuracy of the solutions.

The other important issue is to make sure that boundary conditions and loads should be sufficient so that they lead to a well-posed problem. For example, for the impact problem in the present example, if there is no support of the brittle plate, then it may cause rigid body motions. Without enough boundary conditions, the stiffness matrix may be singular, and subsequently makes the inversion impossible. If the stiffness differences are too large, it will leads to potential numerical difficulties.

10.1.2 Equation Solvers

Almost all numerical simulations you meet in computational engineering will end up with some matrix equations in the form either $\mathbf{KU} = \mathbf{b}$ or more generally $\mathbf{A}\dot{\mathbf{u}} + \mathbf{B}\ddot{\mathbf{u}} + \mathbf{K}\mathbf{u} = \mathbf{b}$, which requires the inversion of large matrices such as $\mathbf{K}$. A direct inversion technique is usually not the best choice. For the problems in computational fluid dynamics, the matrix equations are much more complicated, and often nonlinear. What the solvers essentially do is to solve these matrix equations by an appropriate methods, notably by various iteration methods.

A very important question here is to use the right technique to solve the matrix equation. If you use inappropriate methods, the solution process could be extremely slow, or the program can diverge and crash. It may even give outrageous results, especially for the nonlinear problems. For most finite element packages, they usually suggest a few methods and often have a default iteration type. A sensible choice is to use these recommended methods first before you venture other unfamiliar methods.

In the solving step, the software reads the files generated from the preprocessing step, store them into matrix equations. After solving these equations, they transfer the results, such as $\mathbf{u}$, back into a file or files for postprocessing and visualization. The good things is that this part is virtually hidden, and all you have to do is to finish the task by clicking a few buttons. Most software packages will do these computations automatically. The trouble is that they act like a black box, you can do little or nothing about it if you want to add more functions.

Some scientific packages do allow the user-supplied subroutines or functions to be plugged in their main solvers. However, most of the functions are limited either in the format or the type of problems you can solve. In these situations, your skills in numerical methods and programming become handy in dealing with these problems.

10.1.3 Post-Processing

Once you have run the simulations, you need to use post-processing to visualize your data and explain the results. Post-processing software can have many colourful and fancy visualization functions, you can do many wonderful things such as

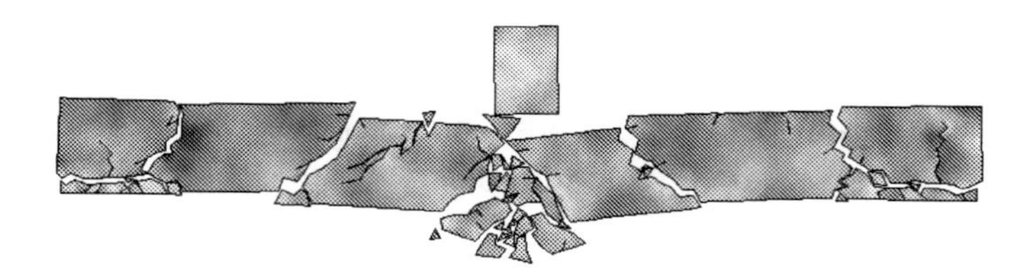

Figure 10.2: Fracture of a concrete beam under impact.

Figure 10.3: Heat transfer on a carbon nanotube array.

viewing the results from various angles and locations, selecting the types of results for displaying, and outputting the visualization into dozens of different formats. Figure 10.2 shows the fracture pattern of a concrete beam due to a drop load where you can see that the post-processing can provide the visualization of stress and fragmentation of each individual piece. Figure 10.3 shows the heat transfer on an array of carbon nanotubes attached to a plate.

The outputs from standard finite element analysis are usually limited to a few popular field quantities such as displacements. The output types depends on the formulation, type of problem and the software you use. For example, the elastic deformation of the beam in this book is displacement-based, and the main output is thus the displacements. If we want to calculate the stresses and other quantities, we have calculate them at post-processing steps. The good news is that almost all finite element packages nowadays allow you to choose the quantities you want to see and they will automatically calculate them for you. If you write your own programs or subroutine plugins, the right choice of integration schemes (usually in terms of Gauss quadrature) and evaluating derivatives are very important. Inappropriate methods will introduce additional and hidden errors for newly calculated quantities.

10.2 Make Sense of Your Results

The most difficult part after post-processing is to explain your results. If the results are what you expected, you might be happy. Do your results make sense? There is a saying or joke: "If a theorist invents a theory, nobody believes it except the theorist himself or herself, if an experimenter discovers something, everybody but the experimenter believes it. However, if a computer scientist or a numerical analyst gets something via simulations, no one trusts the result including the modeller himself or herself." There is some truth behind this because there are so many factors that can affect the results, especially in computer simulations.

In most cases, the first results are not what you expected (wait, there is usually no important new discovery here, unlike Lorentz did in discovering the Chaos!). In these cases, nine of ten chances are that you made some mistakes in either building the numerical model (such as incorrect element types, wrong boundary conditions and/or wrong loads), choosing the wrong parameters, or visualizing your results in a wrong way. The sensible thing to do is to debug and modify your models so that you can get scientifically correct results. This is extremely time-consuming and remember the Ninety-Ninety Rule of Project Scheduling: "The first ninety percent of the task takes ninety percent of the scheduled time. The last ten percent of the task takes the other ninety percent of the time".

Well, you might say, what happens if the results are not what you expected, and you cannot find any mistake in the models. If this is so, you have two choices: 1) if you are a genius, do what Albert Einstein said "If the reality does not fit the theory, then change the reality!"; 2) double check your models (both mathematical equations and numerical models), solution techniques, and ask someone else help you. After going through these steps, you still get the unexpected results, then you either say the software is not good enough, or you have really made some important discovery.

References

Abramowitz M. and Stegun I. A., *Handbook of Mathematical Functions*, Dover Publication, (1965).

Adamatzky A., Teuscher C., *From Utopian to Genuine Unconventional Computers*, Luniver Press, (2006).

Arfken G., *Mathematical Methods for Physicists*, Academic Press, (1985).

Ashby M. F. and Jones D. R., *Engineering Materials*, Pergamon Press, (1980).

Atluri S. N., *Methods of Computer Modeling in Engineering and the Sciences*, Vol. I, Tech Science Press, (2005).

Bathe K. J., *Finite Element Procedures in Engineering Analysis*, Prentice-Hall, (1982).

Carrrier G. F. and Pearson C. E., *Partial Differential Equations: Theory and Technique*, 2nd Edition, Academic Press, 1988.

Carslaw, H. S., Jaeger, *Conduction of Heat in Solids*, 2nd Edition, Oxford University Press, (1986).

Courant R. and Hilbert, D., *Methods of Mathematical Physics*, 2 volumes, Wiley-Interscience, New York, (1962).

Crank J., *Mathematics of Diffusion*, Clarendon Press, Oxford, (1970).

Cook R. D., *Finite Element Modelling For Stress Analysis*, Wiley & Sons, (1995).

Devaney R. L., *An Introduction to Chaotic Dynamical Systems*, Redwood, (1989).

Drew, D. A., Mathematical modelling of two-phase flow, *A. Rev. Fluid Mech.*, **15**, 261-291 (1983).

Fenner R. T., *Engineering Elasticity*, Ellis Horwood Ltd, (1986).

Fowler A. C., *Mathematical Models in the Applied Sciences*, Cambridge University Press, (1997).

Farlow S. J., *Partial Differential Equations for Scientists and Engineers*, Dover Publications, (1993).

Fletcher, C. A. J., Fletcher C. A., *Computational Techniques for Fluid Dynamics*, Vol. I, Springer-Verlag, GmbH, (1997).

Gardiner C. W., *Handbook of Stochastic Methods*, Springer, (2004).

Gershenfeld N., *The Nature of Mathematical Modeling*, Cambridge University Press, (1998).

Goldreich P. and S. Tremaine, The dynamics of planetary rings, *Ann. Rev. Astron. Astrophys.*, **20**, 249-83 (1982).

Goodman R., *Teach Yourself Statistics*, London, (1957).

Gleick J., *Chaos: Making a New Science*, Penguin, (1988).

Hinch E.J., *Perturbation Methods*, Cambridge Univ. Press, (1991).

Hull J. C., *Options, Futures and Other Derivatives*, Prentice-Hall, 3rd Edition, (1997).

Jeffrey A., *Advanced Engineering Mathematics*, Academic Press, (2002).

John F., *Partial Differential Equations*, Springer, New York, (1971).

Jouini E., Cvitanic J. and Musiela M., *Handbook in Mathematical Finance*, Cambridge Univ Press, (2001).

Kardestruncer H. and Norrie D. H., *Finite Element Handbook*, McGraw-Hill, (1987).

Keener J., Sneyd J., *A Mathematical Physiology*, Springer-Verlag, New York, (2001).

Korn G. A. and Korn T. M., *Mathematical Handbook for Scientists and Engineers*, Dover Publication, (1961).

Kreyszig E., *Advanced Engineering Mathematics*, 6th Edition, Wiley & Sons, New York, (1988).

Kant T., *Finite Elements in Computational Mechanics*, Vols. I/II, Pergamon Press, Oxford, (1985).

Langtangen, H P, *Computational Partial Differential Equations: Numerical Methods and Diffpack Programming*, Springer, (1999).

LeVeque R. J., *Finite Volume Methods for Hyperbolic Problems*, Cambridge University Press, (2002).

Lewis R. W., Morgan K., Thomas H., Seetharamu S. K., *The Finite Element Method in Heat Transfer Analysis*, Wiley & Sons, (1996).

Lewis R. W., Gethin D. T., Yang X. S., Rowe R. C., A combined finite-discrete element method for simulating pharmaceutical powder tableting, *Int. J. Num. Meth. Eng.*, **62**, 853-869 (2005).

Mitchell A. R. and Griffiths D. F., *Finite Difference Method in Partial Differential Equations*, Wiley & Sons, New York, (1980).

Moler C. B., *Numerical Computing with MATLAB*, SIAM, (2004).

Murray J. D., *Mathematical Biology*, Springer-Verlag, New York, (1998).

Ockendon J., Howison S., Lacey A., and Movchan A., *Applied Partial Differential Equations*, Oxford University Press, (2003).

Pallour J. D. and Meadows D. S., *Complex Variables for Scientists and Engineers*, Macmillan Publishing Co., (1990).

Papoulis A., *Probability and statistics*, Englewood Cliffs, (1990).

Pearson C. E., *Handbook of Applied Mathematics*, 2nd Ed, Van Nostrand Reinhold, New York, (1983).

Press W. H., Teukolsky S. A., Vetterling W. T., Flannery B. P., *Numerical Recipe in C++: The Art of Scientific Computing*, 2nd Edition, Cambridge University Press, (2002).

Puckett E. G., Colella, P., *Finite Difference Methods for Computational Fluid Dynamics*, Cambridge University Press, (2005).

Revil A., Pervasive pressure-solution transfer: a poro-visco-plastic model, *Geophys. Res. Lett.*, **26**, 255-258 (1999).

Revil A., Pervasive pressure solutoin transfer in a quartz sand, *J. Geophys. Res.*, **106**, 8665-8690 (2001).

Riley K. F., Hobson M. P., and Bence S. J., *Mathematical Methods for Physics and Engineering*, 3rd Edition, Cambridge University Press (2006).

Ross S., *A first Course in Probability*, 5th Edition, Prentice-Hall, (1998).

Selby S. M., *Standard Mathematical Tables*, CRC Press, (1974).

Strang G. and Fix G. J., *An Analysis of the Finite Element Method*, Prentice-Hall, Englewood Cliffs, NJ, (1973).

Smith, G. D., *Numerical Solutions of Partial Differential Equations: Finite Differerence Methods*, 3rd ed., Clarendon Press, Oxford, (1985).

Sukumar N., Moran B., Black T., Belytschko T., An element-free Galerkin method for three-dimensional fracture mechanics, *Computational Mech.*, **20**, 170-175 (1997).

Sukumar N., Moran B., and Belytschko T., The natural element method in solid mechanics,*Int. J. Meth. Eng.*, **43**, 839-887 (1998).

Thomee V., *Galerkin Finite Element Methods for Parabolic Problems*, Springer-Verlag, Berlin, (1997).

Weisstein E. W., http://mathworld.wolfram.com

Wikipedia, http://en.wikipedia.com

Wylie C. R., *Advanced Engineering Mathematics*, Tokyo, (1972).

Versteeg H. K, Malalasekra W., Malalasekra W., *An Introduction to Computational Fluid Dynamics: The Finite Volume Method*, Prentice Hall, (1995).

Yang X. S., Young Y., Cellular automata, PDEs and pattern formation (Chapter 18), in *Handbook of Bioinspired Algorithms*, edited by Olarius S. and Zomaya A., Chapman & Hall / CRC, (2005).

Yang X. S., *Applied Engineering Mathematics*, Cambridge Int. Science Publishing, (2007).

Zienkiewicz O C and Taylor R L, *The Finite Element Method*, vol. I/II, McGraw-Hill, 4th Edition, (1991).

Appendix A

Computer Programs

A.1 Pattern Formation

This simple program solves a nonlinear reaction-diffusion equation and generate beautiful patterns. It is implemented in Matlab using the standard finite difference method.

pattern.m

```
% ---------------------------------------------------
% Pattern formation:  a 15 line matlab program
% PDE form: u_t=D*(u_{xx}+u_{yy})+gamma*q(u)
% where q(u)='u.*(1-u)';
% The solution of this PDE is obtained by the
% finite difference method, assuming dx=dy=dt=1.
% ---------------------------------------------------
% Written by X S Yang (Cambridge University)
% Usage: pattern(100)    or   simply >pattern
% ---------------------------------------------------

function pattern(n)                          % line 1
% Input number of time steps
if nargin<1, n=200; end                      % line 2

% ---------------------------------------------------
% Initialize parameters
% ---- time=100, D=0.2; gamma=0.5; --------------
time=100; D=0.2; gamma=0.5;                  % line 3
```

```
% ---- Set initial values of u randomly ---------
u=rand(n,n);  grad=u*0;                    % line 4

% Vectorisation/index for u(i,j) and the loop ---
I = 2:n-1; J = 2:n-1;                      % line 5

% -----------------------------------------------
% ---- Time stepping ----------------------------
for step=1:time,                           % line 6
% Laplace gradient of the equation         % line 7
 grad(I,J)= u(I,J-1)+u(I,J+1)+u(I-1,J)+u(I+1,J);
 u =(1-4*D)*u+D*grad+gamma*u.*(1-u);       % line 8
% ----- Show results ----------------------------
 pcolor(u);  shading interp;               % line 9
% ----- Coloring and showing colorbar -----------
 colorbar; colormap jet;                   % line 10
 drawnow;                                  % line 11
end                                        % line 12

% ----- Topology of the final surface ----------
surf(u);                                   % line 13
shading interp;                            % line 14
view([-25 70]);                            % line 15
% ------------- End of Program -----------------
```

A.2 Elasticity

This Matlab program provides a simple and yet complete finite element package for simulating elastic deformation of a beam with a given thickness. It first generates 2D triangular mesh, then assembles the finite elements and applies the boundary conditions. The displacements are then displayed as contours in colours.

elasticity.m

```
% -----------------------------------------------
% Solving a linear elastic beam bending problem
% by using the finite element method
```

```
% ------------------------------------------------
% Program by X S Yang (Cambridge University)
% ------------------------------------------------
% After launching Matlab, please type>elasticity;
% Usage: >elasticity(ni, nj);
%   e.g. >elasticity(20,10);
% ------------------------------------------------

function elasticity(ni,nj)

% ------------------------------------------------
% check number of inputs or use default values
if nargin<2,
   disp('Usage: elasticity(ni,nj)');
   disp(' e.g., elasticity(15,4)');
end

% Default values for ni=20, nj=10 --------------
% ni=number of division along beam axis ---------
% nj=number of division in transverse direction -
if nargin<1, ni=20; nj=10; end

% ------------------------------------------------
% ----- Preprocessing  (build an FE model)  -----
% ------------------------------------------------
% Initializing the parameters and beam geometry
% Triangular mesh/elements for 2-D plane stress
% Forces at the top end in units of N (Newton)
fx=1000; fy=50;
% Applied/fixed displacements
u0=0.0; v0=0.0;

% Size of the rectangular beam
% tunit=thickness
height=1; width=0.2; tunit=0.1;

% number of nodes (n) and
% number of elements (m)
n=ni*nj; m=2*(ni-1)*(nj-1);
dx=width/(nj-1); dy=height/(ni-1);

% Initialize the matrices
```

```
A=zeros(2*n,2*n); f=zeros(1,2*n)'; F=f;

% ------------------------------------------------
% Young's modulus Young's modulus =10^8=100 MPa
% This is a very soft material, real materials
% have E=10MPa (rubber) to 1000GPa (diamond)
% We use small Y to get larger displacements
% Poisson ratio nu=0.25
% ------------------------------------------------
Y=10^8; nu=0.25;

% Penalty factor pen for the application of the
% essential boundary conditions
pen=1000*Y;

% Show the results by multiplying these factors
% =1(true displacement); >1 (exaggerated display)
ushow=1; vshow=1;

% ------------------------------------------------
% Show the mesh and grid
subplot(1,3,1);
pcolor(zeros(ni,nj));
set(gca,'fontsize',14,'fontweight','bold');
title('Mesh Grid');

% ------------------------------------------------
% Dividing the whole domain into small blocks
% ------------------------------------------------
% Generating nodes on a rectangular region
k=0;
for i=1:ni,
   for j=1:nj,
      k=k+1;
      x(k)=(j-1)*dx; y(k)=(i-1)*dy;
      node(k)=k;
      xp(i,j)=x(k); yp(i,j)=y(k);
   end
end

% ------------------------------------------------
% Elements connectivity E(1,:), E(2,:), E(3,:)
```

```
k=-1; nk=0;
bcsize=2*(nj+ni-2);

for i=1:ni-1,
      for j=1:nj-1,
      k=k+2;
      nk=node((i-1)*nj+j);
      E(1,k)=nk;
      E(2,k)=nk+1;
      E(3,k)=nj+nk;
      E(1,k+1)=nk+1;
      E(2,k+1)=nj+nk+1;
      E(3,k+1)=nj+nk;
      end
end
% ----------------------------------------------
% Preparing boundary conditions and loads
% Boundary conditions at the bottom (Fixed)
BC(1:nj)=1:nj;
TOPFORCE(1)=[(ni-1)*nj+1];

% ----------------------------------------------
% Assembly to Ku=f (or Au =f in this program)
% Element-by-element assembly

area=height*width/m; s=1/(2*area);
G=Y*[1 -nu 0;-nu 1 0;0 0 (1-nu)/2];
for i=1:m,
   ja=node(E(1,i)); x1=x(ja); y1=y(ja);
   jb=node(E(2,i)); x2=x(jb); y2=y(jb);
   jc=node(E(3,i)); x3=x(jc); y3=y(jc);

%Index matrix
%ID=[i1,i1......i1; i2,i2....i2;
%      j1....j1;j2....j2;k1...k1;k2...k2];
%JD=ID'; A(i,j)=A(ID,JD);
a=...
[2*(ja-1)+1 2*ja 2*(jb-1)+1 2*jb 2*(jc-1)+1 2*jc];
aj=[a;a;a;a;a;a]; ai=aj';
B=s*[y2-y3 0  y3-y1 0   y1-y2  0;
      0  x3-x2  0  x1-x3  0   x2-x1;
      x3-x2 y2-y3 x1-x3 y3-y1 x2-x1 y1-y2];
```

```
% ------------------------------------------------
% Material properties E,\nu are included in G,K
% Stiffness matrix K
   K=B'*G*B*tunit*area;

   for ii=1:6,
      for jj=1:6,
        A(ai(ii,jj),aj(ii,jj))=...
           K(ii,jj)+A(ai(ii,jj),aj(ii,jj));
      end
   end
end

% ------------------------------------------------
% Application of boundary conditions and loads
% ------------------------------------------------
% Essential boundary conditions (u_i=\bar u_i)
for i=1:size(BC,2),
      ij=2*BC(i);
      A(ij-1,ij-1)=A(ij-1,ij-1)+pen;
      F(ij-1)=f(ij-1);
      f(ij-1)=f(ij-1)+pen*u0;
      A(ij,ij)=A(ij,ij)+pen;
      F(ij)=f(ij); f(ij)=f(ij)+pen*v0;
      text(i-0.1,1,'\Delta');
end

% ----- Application of loads --------------------
for i=1:size(TOPFORCE,2),
   ij=2*TOPFORCE(i);
   f(ij-1)=f(ij-1)+fx;
   F(ij-1)=f(ij-1);
   f(ij)=f(ij)+fy;
   F(ij)=f(ij);
   text(0.5,ni,'\rightarrow',...
          'fontsize',20,'color','r');
end
% ------------End of Preprocessing -------------

% ------------------------------------------------
% ----- Solving the equation/problem ------------
```

```
% Solving the equation by simple inversion
u=A\f;

% -----------------------------------------------
% ----- Postprocessing and visualisation --------
% -----------------------------------------------

% Preparing and decomposing the results
k=-1;
for i=1:ni,
   for j=1:nj,
      k=k+2;
      U(i,j)=u(k);
      V(i,j)=u(k+1);
      Fx(i,j)=F(k);
      Fy(i,j)=F(k+1);
   end
end

% -----------------------------------------------
% Visualising the displacements using pcolor() --

subplot(1,3,2);
pcolor(xp+U*ushow,yp+V*vshow,U);
shading interp; colorbar;
set(gca,'fontsize',14,'fontweight','bold');
axis off;
title('U displacement');
subplot(1,3,3);
pcolor(xp+U*ushow,yp+V*vshow,V);
axis off;
shading interp;
set(gca,'fontsize',14,'fontweight','bold');
colorbar;
title('V displacement');
% -------End of program ------------------------
% -----------------------------------------------
```

A.3 Travelling Wave

The finite element method for the wave equation in 1-D case is implemented in the following Matlab program:

Fem_wave.m

```
% ----------------------------------------------
% Solving the 1-D wave equation using the
% finite element method, implemented in Matlab
% written by X S Yang (Cambridge University)
% PDE form: u_{tt}-c^2 u_{xx}=0;   c=speed=1;
% ----------------------------------------------
%n=number of nodes, N=time-step
n=100;

% ----- Initializing various parameters ---------
L=1.0;               % length of domain
speed=1.0;           % wave speed
m=n-1;               % number of elements
time=1;              % total time of simulations

% ----- Time steps and element size ------------
dt=L/(n*speed);  hh=L/m;
N=time/dt;           % Number of time steps

% ----- Preprocessing ---------------------------
% Split the domain into regularly-spaced n nodes
for i=1:n,
          x(i)=(i-1)*L/m;
end
x(1)=0; x(n)=L;

% ----- Finding the element connectivity --------
% Simple 1-D 2-node elements E(1,:) and E(2,:)
for i=1:m,
   E(1,i)=i;
   E(2,i)=i+1;
   h(i)=abs(x(E(2,i))-x(E(1,i)));
end

% ----- Initialization of arrays/matrices --------
```

```
u=zeros(1,n)'; f=zeros(1,n)';
k=[1 -1;-1 1];
A=zeros(n,n);  M=zeros(n,n);

% ----- Element-by-element assembly -------------
% M d^2U/dt^2+KU=f;
for i=1:m,
   A(i,i)=A(i,i)+1/h(i);
   A(i,i+1)=A(i,i+1)-1/h(i);
   A(i+1,i)=A(i+1,i)-1/h(i);
   A(i+1,i+1)=A(i+1,i+1)+1/h(i);
end

% ----- Application of boundary conditions ------
% Fixed boundary  at both ends: u(0)=u(1)=0
A(n,1)=1; A(n,n-1)=0; A(1,1)=1; A(1,2)=0;

% ----- General mass matrix M -------------------
for i=2:n-1,
    M(i,i)=hh;
end;
M(1,1)=hh/2;
M(n,n)=hh/2; Minv=inv(M);

% ----- Preparing time stepping -----------------
D=2*eye(n,n)-dt*dt*Minv*A;

% ----- Initial waveforms with two peaks --------
u0=exp(-(40*(x-L/2)).^2)...
            +0.5*exp(-(40*(x-L/4)).^2);
v=u0'; U=v;

% ----- Plot out the initial waveforms ----------
plot(x,u0); axis([0 L -1 1]);
set(gca,'fontsize',14,'fontweight','bold');
title('Travelling Waves');
xlabel('x'); ylabel('u(x,t)');
axis tight;
set(gca,'nextplot','replacechildren');

% ---------------------------------------------
% ---- Solving the matrix equation -------------
```

```
% ---- Start time-stepping ----------------------
for t=1:N,
        u=D*U-v;
        v=U;            % stored to be used later
        U=u;            % stored as previous values

% ----- Display the travelling wave -------------
% ----- Postprocessing --------------------------
    plot(x,u,x,u0,'-.','linewidth',2);
    axis([0 L -1 1]);
    V(t)=getframe;    % for animation/movie
end

% ----- Replay the animation --------------------
movie(V);
% ----------------------------------------------
```

Index

Lightning Source UK Ltd.
Milton Keynes UK
UKOW04f1017281113

222012UK00018B/558/A

9 781905 986088